Granular Video Computing

with Rough Sets, Deep Learning and in IoT

Granular Video Computing

with Rough Sets, Deep Learning and in IoT

Debarati B Chakraborty
Indian Institute of Technology Jodhpur, India

Sankar K Pal
Indian Statistical Institute Kolkata, India

W€ World Scientific

NEW JERSEY · LONDON · SINGAPORE · BEIJING · SHANGHAI · HONG KONG · TAIPEI · CHENNAI · TOKYO

Published by

World Scientific Publishing Co. Pte. Ltd.
5 Toh Tuck Link, Singapore 596224
USA office: 27 Warren Street, Suite 401-402, Hackensack, NJ 07601
UK office: 57 Shelton Street, Covent Garden, London WC2H 9HE

Library of Congress Cataloging-in-Publication Data
Names: Chakraborty, Debarati B, author. | Pal, Sankar K., author.
Title: Granular video computing : with rough sets, deep learning and in IoT /
    Debarati B Chakraborty, Indian Institute of Technology, Jodhpur, India,
    Sankar K Pal, Indian Statistical Institute, Kolkata, India.
Description: Singapore ; New Jersey : World Scientific Publishing Co. Pte. Ltd., [2021] |
    Includes bibliographical references and index.
Identifiers: LCCN 2020040374 | ISBN 9789811227110 (hardcover) |
    ISBN 9789811227127 (ebook) | ISBN 9789811227134 (ebook other)
Subjects: LCSH: Computer vision. | Automatic tracking.
Classification: LCC TA1634 .C435 2021 | DDC 006.7--dc23
LC record available at https://lccn.loc.gov/2020040374

British Library Cataloguing-in-Publication Data
A catalogue record for this book is available from the British Library.

For any available supplementary material, please visit
https://www.worldscientific.com/worldscibooks/10.1142/12013#t=suppl

Desk Editors: George Vasu/Amanda Yun

Typeset by Stallion Press
Email: enquiries@stallionpress.com

Dedicated to our Parents

Preface

The volume *Granular Video Computing with Rough Sets, Deep Learning and in IoT* is an outcome of the investigations initiated by the authors in 2011 as a part of granular computing research at the Center for Soft Computing Research (CSCR), a national facility, at the Indian Statistical Institute (ISI), Kolkata. The center was established in 2005 by the Department of Science and Technology, Govt. of India (with Prof. Sankar K. Pal as the Principal Investigator) under its prestigious IRHPA (Intensification of Research in High Priority Areas) program. It is now an Affiliated Institute of ISI since 2010.

Video computing is one of the basic steps in human–computer interactive (HCI) systems. Its primary applications lie in automated surveillance, gesture recognition, event detection, behavior recognition and internet of things (IoT). The steps of video computing involve tracking of moving objects, motion analysis and video interpretation. There are several unsolved issues in the task of video computing. These include handling uncertainties arising from different types of ambiguities in video sequences. Ambiguity may result, for example, from the absence of complete information regarding the number of moving object(s), background, newly appeared object(s), moving-to-static object(s), partly moving object(s), partially occluded or overlapped object(s) and fully occluded or overlapped moving object(s). Accuracy and speed in object recognition and video understanding, and performing these tasks in the absence of large labeled dataset are some other issues of concern.

Granular computing (GrC) is a nature inspired information processing paradigm whose two major components are: formation of granules and performing computation with granules. When a problem involves incomplete, uncertain and vague information, it may be difficult to differentiate distinct elements and one may find it convenient to consider granules for its handling. Granules are composed of elements or objects that are drawn together by, say, indiscernibility, similarity and/or functionality. They evolve in the process of abstraction and derivation of knowledge from data. Each granule according to its shape and size, and with a certain level of granularity, may reflect a specific aspect of the problem. Accordingly, GrC became an effective framework in the design and implementation of efficient and intelligent information processing systems for various real-life decision-making applications. Since GrC deals with clump of indiscernible data together, rather than individual data points, it leads to computation gain. This makes it suitable for mining large datasets.

Problems in video computing often involve incomplete, uncertain and vague information and it is sometimes difficult to differentiate distinct elements. Therefore, consideration of GrC appears to be effective here both in terms of performance and computation time. Judiciously forming granules with the spatial as well as temporal information of video sequences may characterize a specific aspect of the problem or task in video analysis. Further, the requirement of video understanding in IoT is increasing day by day. Realization of such a granular video computing model in IoT framework is, therefore, a need of today's society. Theory of rough sets provides a popular mathematical framework for GrC. The focus of the theory is on the ambiguity caused by limited discernibility of objects in the domain of discourse. Its key concepts are those of object 'indiscernibility' and 'set approximation'. Two major characteristics of the theory that have drawn the attention of researchers in AI and data science are uncertainty handling (using lower and upper approximations) and GrC (using information granules). Lower and upper approximate regions of a set, respectively, denote the granules which definitely and definitely and possibly belong to it. These characteristics made the theory widely applicable in tasks like pattern recognition and machine learning, feature reduction/selection, image processing, data mining and knowledge discovery. This theory is also proven to be useful in incomplete knowledge base.

Deep learning architecture is a popular tool in the problem of object recognition. The task of labeling different objects in an image with the correct class and predicting the bounding boxes along with probability are done with this network. While deep learning is a computationally intensive process, and the aforesaid granular computing paradigm, on the other hand, leads to gain in computation time, it may be appropriate and logical to make their judicious integration so as to make the deep learning framework efficient in terms of computation time.

This book, *Granular Video Computing with Rough Sets, Deep Learning and in IoT*, provides a treatise in a unified framework describing how the aforesaid challenging issues in video computing can be addressed with rough set theoretic granular computing framework. Three major issues dealt with, among others, include: uncertainty handling, reducing the computation time, and performing in unsupervised mode. Based on the existing as well as new results, the book is structured according to major challenges and tasks in video computing with a balanced mixture of theory, algorithm and applications. Tasks considered are: tracking in both supervised and unsupervised frameworks, measuring trustability of tracking, object recognition from scene with linguistic description, and video understanding. Other important tasks considered are: formation of deep learning network incorporating GrC for video analysis, and experimental demonstration of the effectiveness of GrC-based video event recognition in IoT. Several new terms/tools were evolved in the process. These include granular flow graph, neighborhood rough filter, intuitionistic entropy, motion granules, motion entropy, video conceptualization, and granulated CNN (convolutional neural network). Here, defining different neighborhood granulated domains over the video frames, in spatial/temporal space, and extracting the meaningful rough lower and upper approximated regions are crucial. Investigations developing all these methodologies, theories and algorithms, along with extensive experimental results on different kinds of video sequences sensed by RGB, Kinect and IR sensors, and the associated novelties are summarized in seven chapters of the book.

Chapter 1 provides an introduction to video computing, granular computing, rough sets, deep learning and IoT for the convenience of readers. These are followed by a description about the scope of the book. The primary problem of video processing, namely, object tracking, is addressed in Chapters 2 to 5 with different degrees of complexity. Chapters 6 and 7 deal with the problems of object recognition from a scene and video conceptualization, respectively.

Chapter 2 describes the task of partially supervised tracking of single moving object based on object–background spatio-temporal segmentation. Here rough entropy is used for optimum spatial segmentation using both equal sized and unequal sized spatio-color rectangular granules, whereas 3-point background estimation is used for temporal segmentation. Chapters 3 and 4 both focus on unsupervised tracking of multiple moving objects in different directions and in different speeds. They function over arbitrary shaped neighborhood granules. In Chapter 3, merits of granular rule-based classification are explored for object–background classification. Parameters (antecedents) of the rule-base are obtained from the 2-D spatio-color and 3-D spatio-temporal neighborhood granules, as defined over the lower approximated frame of moving objects in temporal domain. The rule-base is made adaptive so as to be able to track a newly appearing object while tracking others. Granular flow graph enables intelligent updating of rule-based parameters. Tracking in Chapter 4, on the other hand, is based on prediction of object locations. Here a rough filter is designed whose output provides initial labeling of moving objects in terms of their lower-upper approximations in spatio-color granular space. Prediction in a frame involves computation of velocity granules and acceleration granules, and 'intuitionistic entropy' using filter outputs (estimations) on previous frames. Minimizing this entropy over ambiguous pixels enables tracking multiple-objects even with complete occlusion and overlapping. Chapter 5 addresses the issue of trust ability of tracking systems. Here, formulations of various indices to evaluate the performance of tracking are explained with experimental validation.

The problem of object recognition from a scene has been addressed in Chapter 6. Here, a granulated deep CNN is designed for the first time, incorporating granulation in the convolution layer, to reduce the computation time significantly with some compromise in performance. Z-numbers are used to provide linguistic description of the scene based on its recognized static and moving objects. This is unique. A new application toward video understanding, namely, video conceptualization is described in Chapter 7. It first involves granulation of the video sequence in terms of motion granules and formation of rough object–background set. Motion entropy computed on this characterizes the uncertainty present in different moving objects, and classifies them as either random movement (noise), continuous (or predictable) movement, or sudden-changed (or unpredictable) movement. This is how one can conceptualize the overall content of the video, and identify, say, the frames containing most information for further analysis. As a real-life application, the significance of this task is demonstrated in an IoT set up with physical devices on online videos in identifying the frames containing unusual movements.

Most of the texts presented in this volume are from our published research work. The related and relevant existing approaches or techniques are included wherever necessary. Directions for future research in the concerned topic are provided in each chapter. A comprehensive bibliography on the subject is also appended in every chapter. References to some of the related investigations might have been omitted because of oversight or ignorance.

The book is unique in its character, and will be useful to graduate students and researchers in computer science, electrical engineering, system science, data science and information technology, both as a textbook and a reference book for some parts of the curriculum. The researchers and practitioners in industry and R&D laboratories working in the fields of system design, pattern recognition, machine learning, deep neural networks, Big data analytics, video surveillance, IoT and soft computing or computational intelligence will also be benefited.

We take this opportunity to acknowledge Prof. Andrzej Skowron, Warsaw University, Poland, for his encouragement and support in the endeavor. We owe a vote of thanks to Dr. Amanda Yun of World Scientific, Singapore, for co-ordinating the project, as well as the office staff of our Soft Computing Research Center for their support. The book was written when Prof. S.K. Pal held a Distinguished Professorial Chair at the Indian National Science Academy (INSA).

About the Authors

Debarati B. Chakraborty is currently a young faculty associate in the Department of Computer Science and Engineering at Indian Institute of technology, Jodhpur, India. She has done her Ph.D. from Jadavpur University, Kolkata, India in 2017, M. Tech from National Institute of Technology, Rourkela in 2010 and B. Tech from West Bengal University of Technology in 2008. She has conducted her research work towards Ph.D. at the Center for Soft Computing Research in Indian Statistical Institute during 2011–2017. She carried out her post doctoral research from the same institute during 2017–2019.

Her research interests include video processing, rough sets, granular computing, soft computing, artificial intelligence, machine learning, deep learning, and IoT. She received the prestigious young scientist award by Indian Science Congress Association in 2012.

Sankar K. Pal is currently a National Science Chair, SERB-DST, Govt. of India. He is a Distinguished Scientist and former Director of Indian Statistical Institute, a former Distinguished Professor of Indian National Science Academy, and a former Chair Professor of Indian National Academy of Engineering. He founded the Machine Intelligence Unit and the Center for Soft Computing Research: A National Facility in the Institute in Calcutta. He received a Ph.D. in Radio Physics and Electronics from the University of Calcutta in 1979, and another Ph.D. in Electrical Engineering along with DIC from Imperial College, University of London in 1982.

He worked at the University of California, Berkeley and the University of Maryland, College Park in 1986–1987; the NASA Johnson Space Center, Houston, Texas in 1990–1992 & 1994; and in US Naval Research Laboratory, Washington DC in 2004. Since 1997 he is a Distinguished Visitor of IEEE Computer Society (USA) for the Asia-Pacific Region, and held several visiting positions in Italy, Poland, Hong Kong and Australian universities.

Pal is a Fellow of IEEE, the World Academy of Sciences (TWAS), International Association for Pattern recognition, International Association of Fuzzy Systems, International Rough Set Society, and all the four National Academies for Science/Engineering in India. He is a coauthor of twenty books and more than four hundred research publications in the areas of Pattern Recognition and Machine Learning, Image Processing, Data Mining and Web Intelligence, Soft Computing, Neural Nets, Genetic Algorithms, Fuzzy Sets, Rough Sets, Cognitive Machine and Bioinformatics. He introduced and promoted the soft computing research & teaching in India. He visited forty five countries as a Keynote/Invited speaker or an academic visitor.

He received the 1990 S.S. Bhatnagar Prize (which is the most coveted award for a scientist in India), 2013 Padma Shri (one of the highest civilian awards) by the President of India, and many prestigious awards in India and abroad including the 2000 Khwarizmi International Award from the President of Iran, 2000–2001, 1993 NASA Tech Brief Award (USA), 1994 IEEE Trans. Neural Networks Outstanding

Paper Award, 1995 NASA Patent Application Award (USA), 1999 G.D. Birla Award, 1998 Om Bhasin Award, 2005–2006 Indian Science Congress-P.C. Mahalanobis Birth Centenary Gold Medal from the Prime Minister of India for Lifetime Achievement, 2007 Sir J.C. Bose National Fellowship, 2015 DAE Raja Ramanna Fellowship, 2015 INAE-S.N. Mitra Award, 2017 INSA-Jawaharlal Nehru Birth Centenary Lecture award, 2018 INSA Distinguished Professorial Chair, and 2020 National Science Chair, Govt. of India.

Pal acts(ted) an Associate Editor of *IEEE Trans.* PAMI (2002–2006), *IEEE Trans. NN* (1994–1998 & 2003–2006), *Neurocomputing* (1995–2005), *Pattern Recog. Lett.* (1993–2011), *Int. J. Patt. Recog. & Art. Intell., Applied Intelligence, Inform. Sci., Fuzzy Sets and Syst., Fundamenta Informaticae, LNCS Trans. Rough Sets, Int. J. Comput. Intell. and Appl., IET Image Process.* (2007–2019), *Ingeniería y Ciencia, and J. Intell. Inform. Syst.*; Editor-in-Chief, *Int. J. Signal Processing, Image Processing and Pattern Recognition*; a Book Series Editor, Frontiers in Artificial Intelligence and Applications, IOS Press, and Statistical Science and Interdisciplinary Research, World Scientific; a Member, Executive Advisory Editorial Board, *IEEE Trans. Fuzzy Systems, Int. Journal on Image and Graphics, Int. Journal of Approximate Reasoning*, and Data-Centric Engineering (Cambridge Univ.); and a Guest Editor of IEEE Computer, *IEEE Trans. SMC*, and Theoretical Computer Science.

Contents

List of Figures

List of Tables

List of Algorithms

Chapter 1

Introduction: Video Processing, Granular Computing, Rough Sets, Deep Learning and IoT

1.1 Introduction

Video processing is one of the basic steps in human–computer interactive (HCI) systems. Its primary applications lie in automated surveillance, gesture recognition, event detection, behavior recognition, and internet of things (IoT), among others. The steps of video processing primarily involve tracking of moving objects, motion analysis and video interpretation.

The problem of tracking is the main part of video computing and it has been studied over decades (Yilmaz *et al.*, 2006; Maggio and Cavallaro, 2010; Smeulders *et al.*, 2014; Wu *et al.*, 2015). There are a number of methods for supervised, partially supervised (with initial manual interactions) (Wang *et al.*, 2012; Milan *et al.*, 2014; Zhang *et al.*, 2014; Pernici and Bimbo, 2014; Bai *et al.*, 2015; Kwon and Lee, 2014), and unsupervised (without manual interactions, mostly based on background estimation) (Dey and Kundu, 2013; Henriques *et al.*, 2015; Chi *et al.*, 2017) tracking. One may note that in video tracking the complete information on object–background is not always available. This makes the prediction difficult as there exist several uncertainties arising from, e.g., changes in shapes/sizes of moving object(s), changes in direction and motion of the object(s), and changes in the number of object(s), sudden appearance of object,

object overlapping, and occurrence of total/partial occlusion to background. In the existing pattern recognition literature there are several methods to deal with ambiguities and uncertainties using, for example, statistical and probabilistic methods (Stauffer and Grimson, 1999a, 2000), fuzzy sets and fuzzy logic (Fang *et al.*, 2006), and kernel-based methods (Comaniciu *et al.*, 2003; Shen *et al.*, 2010). Over the years, researchers have been trying to improve the accuracy and speed both in detection and tracking. These methods can be broadly divided into two types. In one kind of approach, the prior knowledge about the number, size, shape and appearance of objects is required (Yilmaz *et al.*, 2006). The other approach is based on background estimation (Cheung and Kamath, 2004; Stauffer and Grimson, 1999b; Cucchiara *et al.*, 2003; Maddalena and Petrosino, 2008) which, in turn, needs several input frames available to approximate the background model. But in the practical scenario, prior knowledge may not always be available. Also there may not be enough input frames available to estimate the exact background. Besides, the cases like overlapping and occlusion may not be handled with background estimation. This book discusses some of the benchmark methods along with the ones developed by the authors dealing with those issues.

The trustability of tracking algorithms is another concern that needs to be well defined during their real-life implementation. A tracking algorithm could be effective in different scenarios to different degrees and sometimes may not be even effective at all. Therefore, quantifying the performance of an algorithm for different environments even for the same task is a crucial issue. This type of quantification helps a system to chose the most suitable solution with respect to an application. In other words, the quantification enhances the trustability or reliability of any video tracking-based AI system. The issue of quantifying the performance of any tracking algorithm has been addressed for over more than a decade. There exist several indices to evaluate the performance of tracking (Black *et al.*, 2003; Kasturi *et al.*, 2009; Nawaz *et al.*, 2014; Čehovin *et al.*, 2016; Fang *et al.*, 2017). In this book, we have described several new and existing such measures. Experimentally measured trustability values of some benchmark tracking algorithms under different situations are also shown.

Object recognition from scene is another major concern of video computing. Code-book formation (Fergus *et al.*, 2003; Leibe *et al.*, 2008), combination of different classifiers (Zhang *et al.*, 2007), part-based model representation (Felzenszwalb *et al.*, 2010) are a few popular approaches, among others, for object recognition from scene. But all these approaches require high computation time even during the testing phase while performing this task. Computation time is a crucial issue in video computing as video is a stream data. A feasible solution in this regard is provided with a newly developed deep learning framework.

Understanding the content of a video is the last and final step of video computing. There are several approaches aimed toward it. Activity pattern analysis (Stauffer and Grimson, 2000), discovery of key patterns (Yang *et al.*, 2009), diffusion map framework (Liu *et al.*, 2010), and AND-OR graph for story-line understanding (Gupta *et al.*, 2009) are a few popular approaches for video understanding. But understanding the total content of a video sequence is a time-consuming approach. It also needs huge amount of labeled data and trained network for this task. These issues make the application of video understanding difficult in real-time scenarios and hence in internet of things (IoT). Here a probable solution to those limitations is provided with the results for unsupervised video motion analysis with real-time data acquisition in IoT.

In this volume, a rough set theory is shown as an effective tool for handling the uncertainties generated due to the aforementioned issues in video computing. Theory of rough sets (Pawlak, 1992) has become a popular mathematical framework for image processing problems (Sen and Pal, 2009a; Pal *et al.*, 2005). The focus of the theory is on the ambiguity caused by limited discernibility of objects in the domain of discourse. Its key concepts are those of object 'indiscernibility' and 'set approximation'. The primary use of rough set theory has so far mainly been in generating logical rules for classification and prediction (Skowron and Rauszer, 1992); thereby making it a prospective tool for various tasks including pattern recognition, image processing, feature selection, data mining and knowledge discovery from large datasets. Use of reducts based on rough set rules has significant role for dimensionality reduction/feature selection by discarding redundant features (Swirniaski, 2001); thereby having potential application for mining large datasets (Komorouski *et al.*, 1999). Rough set theory provides intelligent techniques to discover data dependencies, to evaluate the importance of attributes, to discover patterns in data, to reduce redundancy, and to recognize and classify objects (Hassanien *et al.*, 2008; Swirniaski, 2001; Yao, 2011; Peters and Borkowski, 2004). Though the concept of rough set has been applied in several areas of pattern recognition, its use in the problem of video computing has not been addressed adequately.

Rough set is widely used in granular computing. Granulation is a basic step of human cognition system and hence a part of natural computing (Pal and Meher, 2013; Yao *et al.*, 2013). This concept was first introduced by Zadeh (1997) in fuzzy set theoretic framework. According to the concept, granulation involves partitioning of an object into granules, with a granule being a clump of elements drawn together by indistinguishability, equivalence, similarity and functionality. Granular computing is regarded as a label of theories, methodologies, techniques and tools that make use of granules in the process of problem solving. It is a mode of computation in which the objects of computation are granular variables. The two basic issues of granular computing are: formation of granules and computing with granules. Granules evolve in the process of abstraction and the derivation of knowledge from data. Granular computing is expected to make a process faster as the computations are done with the granules instead of considering each data point individually.

This book provides some new theories and methodologies, and applications demonstrating the effectiveness of rough set theoretic granular computing for performing different tasks related to tracking moving object(s) and performing video analysis in a sequence. The tasks that are considered include partially supervised video tracking, unsupervised video tracking, multi-object tracking, occlusion/overlapping handling, measuring trustability of tracking algorithms, object recognition in video scene and video conceptualization. Rough sets, rough rule-base, neighborhood rough sets, flow graph, probability theory, and deep learning are used among others, as paradigms for dealing with uncertainties. The ways of forming various unequal 2-D granules, 3-D granules, spatio-temporal neighborhood granules, two-layered motion granules are described. Development of new hybrid tools, e.g., granular rough rule-base, rough flow graph, neighborhood rough filter over videos, and granulated deep learning are explained. This makes the processing faster and more accurate. These tools are proven to be more effective than the existing ones in dealing with uncertainties. Gray color or R-G-B and R-G-B-D video sequences are considered as input data. We have also provided a new application towards video understanding, namely, video conceptualization along with its successful testing in an architecture of internet of things (IoT) with real-time data acquisition.

Before the motivation and scope of the book is stated in Section 1.7, an overview and preliminaries of video computing, deep learning, IoT and some features of rough set theories and granular computing, as relevant to this discipline, are provided. The basic steps related to video computing along with some state-of-the-art methods and quantitative indices are discussed in Section 1.2. The underlying ideas and issues related to granular computing are mentioned in Section 1.3. The theory of rough sets is explained in Section 1.4. It also includes the definitions of neighborhood rough sets and the use of rough sets in different areas of image processing. The basics of deep learning and its applications in image processing are discussed in Section 1.5. The definition, features and applications of internet of things (IoT) are provided there in Section 1.6.

1.2 Video Computing

A video sequence actually is a collection of still images captured continuously. The issues like availability of large data storage and high quality video cameras, and the increasing requirements for automated video analysis in several fields. make video computing a crucial part of machine intelligence. As a result, video analysis has drawn tremendous attention, and several investigations have been conducted till now. Video analysis can be divided into three basic steps, as follows: tracking of moving objects through out a sequence, recognizing the object(s) present in a scene and analysis of object motion to recognize their behavior. Therefore, the use of video computing is required mainly in the areas involved in IoT, such as: (i) moving object recognition, that is, human identification based on its trajectory, automatic object detection, etc; (ii) surveillance, that is, detection of unusual activities or unlikely events automatically by analyzing video streams generated in that place; (iii) video indexing, that is, annotation and retrieval of videos in multimedia databases automatically; (iv) human–computer interaction, which includes gesture recognition and sign language interpretation; (v) traffic monitoring, by real-time analysis of traffic statistics; and (vi) designing automatic vehicles, with video-based path planning and obstacle avoidance capabilities. Before going into the details of other video processing techniques, object tracking from video images is discussed in the next section.

1.2.1   *Video Tracking*

The task of video tracking is to track moving object(s) from a video sequence. It is the primary and the most important step of video computing. The accuracy of any video computing application is initially dependent on the accuracy of its inline video tracking algorithm. The algorithms so far designed for tracking are simplified with some pre-defined assumptions. For example, almost all tracking algorithms assume that the object motion is smooth with no abrupt changes. Constant velocity or constant acceleration are considered further to model the motion of the moving object(s) in some methods. The information on the total number of moving objects, their shape/size and initially labeled model is also given as input in most of the exiting approaches.

Different types of sensors have been used to track moving objects in static background. Some such examples are, multiple cameras (Huang and Fu, 2011), PTZ camera (Liu *et al.*, 2015), kinect sensor (Shum *et al.*, 2013) and IR sensor (Davis and Sharma, 2007). The basic steps of video tracking are representation of moving objects, feature selection and object tracking. These steps are discussed in the following sections.

1.2.1.1   *Extraction and Modeling of Moving Objects*

Representation of moving objects is a very crucial issue in the task of tracking. Representation means how a machine interprets/views the objects in a method. The rest of the processing towards tracking is conducted assuming the said representative is the moving object. In order to do so, at first, the region of interest is extracted out from the frame and then it is represented with some mathematical modeling technique. The most popular methods to extract out the region of interest are:

- Point based, where the region of interest is a single point, i.e., it is centroid (Veenman *et al.*, 2001), or a collection of points (Serby *et al.*, 2004).
- Geometric shape based, where the region of interest is some geometric shape which contains the object region. The most popular shapes by which object(s) are represented so far have been rectangular (Briechle and Hanebeck, 2001; Comaniciu *et al.*, 2003; Xiao and Oussalah, 2016) and elliptical (Milan *et al.*, 2014).
- Contour or silhouette based, where the region of interest is either the shape (contour) of the object or the region within the shape (silhouette) (Dey and Kundu, 2013).
- Skeletal based, where the region of interest is the skeleton of the object contour. This information may be extracted by applying medial axis transform to the object silhouette (Ali and Aggarwal, 2001) or by using the recently developed kinect sensor (Janoch *et al.*, 2011; Lai *et al.*, 2011).

The extracted regions, labeled as the moving object(s), are then represented in some mathematical form for further computation. The popular approaches of object representation are mainly based on (i) probability density, where the object representation may be parametric, e.g., mixture of Gaussian (Stauffer and Grimson, 1999b), metric-weighted linear representations (Li *et al.*, 2016) or non-parametric, mostly with histograms (Comaniciu *et al.*, 2003); (ii) templates, i.e., the 2-D spatial and color information present in the extracted regions (Briechle and Hanebeck, 2001); (iii) active appearance models, which model the shape and changing appearances of objects simultaneously (Nguyen and Smeulders, 2004); and (iv) 3-D modeling, where the object(s) are modeled in a 3-D space. This can be done by collecting information from multiple cameras, so that a scene can be viewed from different angles and those together will construct the 3-D model (Huang and Fu, 2011), or by accumulating the depth features obtained by a depth sensor, e.g., kinect sensor (Janoch *et al.*, 2011; Lai *et al.*, 2011).

1.2.1.2   *Feature Selection for Tracking*

Selection of the right features for modeling the object/background plays a critical role in the task of tracking. Proper selection of features makes the object of interest easily separable from its surroundings. Feature selection is closely related to object representation. For example, color is used as a feature for histogram-based appearance representation, while for contour-based representation, object edges are usually used as features. Most of the tracking algorithms use a combination of these features. The details of common visual features are as follows.

- Color: The apparent color of a scene is influenced primarily by two physical factors, (1) the spectral power distribution of the illuminant and (2) the surface reflectance properties of the object. In image processing, the RGB (red, green, blue) color space is usually used to represent color. However, the RGB space is not a perceptually uniform color space, that is, the differences between the colors in the RGB space do not correspond to the color differences perceived by humans. Whereas, LUV (lightness universal value) and HSI (hue, saturation, intensity) are perceptually uniform color spaces. As none of the individual color spaces have proved to be more effective, a variety of them have been used.
- Edges: Object boundaries are usually generated based on strong changes in image intensities. Algorithms that track the boundary of the objects usually use edges as the representative feature.
- Optical flow: Optical flow is a displacement vector which defines the translation of each pixel in a region from frame-to-frame. It is mainly computed using the brightness values, assuming that brightness of corresponding pixels in consecutive frames should be constant if no movement occurs. Optical flow is commonly used as a feature in motion-based segmentation and tracking applications.
- Depth: Kinect and IR (infrared) are the two recently developed popular depth sensors. The depth values using kinect or IR sensors are acquired in such a way that the moving objects always have distinct gray level values than that of their corresponding backgrounds. It is obvious that the moving object(s) move in front of their corresponding background, i.e., nearer to the sensor, if they do not get occluded. The IR sensor works based on the reflected light with IR wavelength. So, objects closer to the sensor produce

brighter regions. But, in case of kinect sensors, the objects nearer to the sensors have less depth values resulting in darker regions.

1.2.1.3 *Tracking of Moving Objects*

The task of tracking object(s) from video sequences can be broadly classified into two categories. Those are: (i) object tracking with initially labeled object(s) or partially supervised tracking and (ii) object tracking without any labeling or unsupervised tracking. There exist hundreds of methods for the former, but there are scarcely any for the latter case. We are going to discuss a few of the existing benchmark tracking methods for both the aforesaid types in the following sections.

The basic steps of these two types of tracking techniques are shown in Fig. 1.1. One may note that the initial information of the labeled object(s) is given in the partially supervised techniques, whereas the same is extracted out without manual interactions during the training phase of unsupervised techniques.

1.2.1.4 *Partially Supervised Tracking*

The initial step of all the partially supervised approaches is locating the target regions in the initial frames manually and then tracking those in the upcoming video stream. Most of the existing literature on tracking focused on this task with different degrees of complexity in

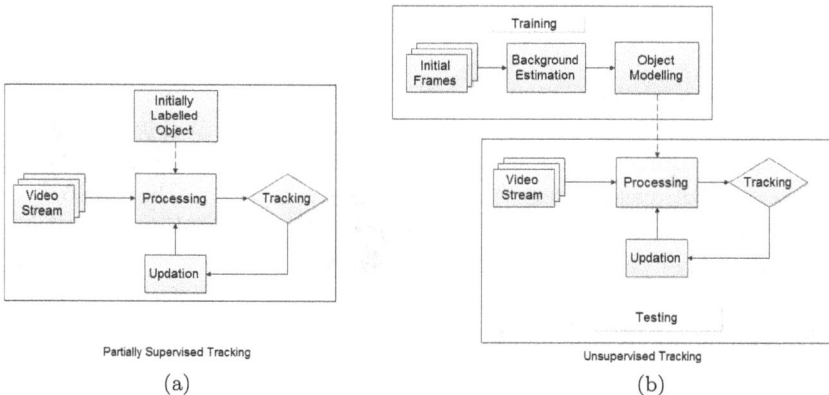

Fig. 1.1: Basic steps of (a) partially supervised tracking and (b) unsupervised tracking.

different video sequences. There are a few unsupervised approaches, too. The underlying concept of unsupervised tracking is discussed in the following section.

1.2.1.5 *Unsupervised Tracking*

The unsupervised tracking techniques can be divided into two phases, namely, training and testing (see Fig. 1.1(b)). The background of a video sequence is estimated from the initial input frames during the training phase and the tracking is done over the rest of the frames during testing. In some cases, instead of background, object model is estimated during the training phase.

Here in the book, we have discussed the problems and methods of mainly partially supervised and unsupervised video tracking. Some popular methods relevant to the specific tasks are discussed in corresponding chapters of the book before describing the recent ones.

1.2.2 *Trustability Measure of Tracking*

There are a number of methods for video tracking, but every method is not equally effective for all the scenarios. Therefore quantification of the trustability of the tracking algorithms is an important issue for real-time applications. There exist several indices to evaluate the performance of tracking in this regard. But most of the indices are defined with ground truth information. That is, these indices measure the similarity between the tracked region and its corresponding ground truth region. The details of the existing indices are provided there in Chapter 5. Apart from those we have described a few recently defined measures to evaluate the performance of tracking. These measures do not require any ground truth or trajectory. The said measures quantify the ambiguities in the tracked regions with respect to that of the object models. Their details are given in Chapter 5.

Let us now discuss the tasks of object recognition from a scene and unsupervised video understanding/conceptualization of videos. These are the other two components of video processing that are addressed in this book apart from tracking.

1.2.3 *Object Recognition from Scene*

Object recognition from scene is an important part of image or video understanding. Different steps of this task include identification, categorization, viewpoint or space estimation, and finally activity recognition of object(s). The primary challenges in the task of object recognition arise due to: different view points of object(s), change in scale and illumination, occlusion, object deformation, intra-class variation, etc.

There are several approaches aimed towards object recognition from a scene (Mottaghi *et al.*, 2016; Gan *et al.*, 2018; Held *et al.*, 2016). Here we discuss a few benchmark methods. Learning a codebook of object features and recognizing new objects from it is one of the popular approaches in object detection. Fergus *et al.* (2003) proposed an unsupervised generative model for configurations of the codebook words of objects. Leibe *et al.* (2008) then came up with a shape model to specify where and for which object(s) the codebook should be used. Zhang *et al.* (2007) investigated combination of different detectors and descriptors with a classifier that can effectively use the information provided by the features. These features are sparse, so even if a generative model is learned, only a sparse set of object points can be generated. Part-based models have been proved successful on difficult object detection datasets. Felzenszwalb *et al.* (2010) proposed a deformable part model for object detection and it has proved to be a successful one. Note that the computation time in testing is high for all the aforementioned approaches. This is a challenging issue while dealing with video sequences. Here comes the significance of granular computing to reduce the computation time. Methods developed in this line are described in Chapter 6.

1.2.4 *Video Understanding*

Understanding the content of a video sequence is the final objective of video computing. The actions that are taken in an automated system are mainly dependent on this step. There are several kinds of analyzing techniques aimed towards it such as scene understanding, behavior understanding, storyline understanding, to mention a few. Chapter 7 describes some methods to analyze the nature of different types of motions present and to detect unpredictable changes in motion in a video sequence. This is done with rough set theoretic granular computing. The method is unsupervised, that is, the estimation is performed without any prior knowledge. This process is named as 'conceptualization'. It means, the video is conceptualized with the motion patterns of objects present in a sequence.

There are plenty of approaches that deal with video understanding (Borges *et al.*, 2013). Brand (1996) first proposed a way to interpret videos of manipulation tasks by using psychology-based causal constraints to detect meaningful changes in motions. Stauffer and Grimson (2000) then came up with an approach of activity pattern analysis by accumulating information from multiple cameras. Yang *et al.* (2009) proposed a method of discovering the key patterns of motion automatically by using a low level feature, pixel-wise optical flow, several of which are embedded later in a diffusion map framework. Liu *et al.* (2010) proposed a hybrid parallel computing framework for video understanding and retrieval based on the Map-Reduce programming model where SVM was used for training. Gupta *et al.* (2009) defined a technique to understand the storyline of a sequence in terms of AND-OR graph which corresponds to causal relationships in terms of spatio-temporal constraints. Zaidenberg *et al.* (2013) described a way of video understanding by people tracking and analyzing the trajectories over a temporal window, and clustered using the Mean-Shift algorithm; this method is then applied for group behavior recognition. A social behavior recognition model was defined by Burgos-Artizzu *et al.* (2012) where the continuous videos are segmented into action bouts by building a temporal context model to combine features from spatio-temporal energy and agent trajectories. Recently, an unsupervised video understanding method has been proposed by Milbich *et al.* (2017) where combinatorial sequence matching is performed to train a CNN, which is carried out with a huge amount of labeled dataset. Another way of video understanding with desktop action recognition from first person's view was developed by Cai *et al.* (2018). It was mainly focused on hand motion analysis. All of the aforementioned approaches either need initial manual intervention or huge amount of labeled dataset for training. However, in real-life applications, one may come up with a solution to make a video sequence grossly interpretable without manual intervention or without training the network with labeled data. The key frames with more information will be extracted out in this process from which the storyline of a video sequence could be further analyzed. Chapter 7 describes such an unsupervised approach to analyze the probable events in video sequence. This is done with the nature of movements present in the sequence.

In this book, the merits of rough set theoretic granular computing are demonstrated to handle the uncertainty present in different tasks of video computing. Therefore, the underlying concepts of granular computing and characteristics of rough theory are discussed in Sections 1.3 and 1.4 for the convenience of the readers. Section 1.4 also discusses image definition in rough set theoretic framework, and rough entropy of an image.

1.3 Granular Computing

Granular computing (GrC) is a nature inspired computing paradigm of information processing. It is concerned with the processing of complex information entities as basic elements, called 'granules of information'. Granules are collections of entities that usually originate at simple object level (data) and are arranged together due to their similarity, functionality or physical adjacency, indistinguishability, coherency, etc. They evolve in the process of abstraction and derivation of knowledge from data. The method of formation and representation of granules is known as granulation.

1.3.1 *Granulation*

Granular computing is more of a theoretical perspective than a coherent set of methods or principles. From a theoretical perspective, it encourages any approach that deals with the problems that recognize and exploit the knowledge present in the data at various levels of resolution or scales. In this sense, it encompasses all methods which provide flexibility and adaptability in the resolution at which knowledge or information is extracted and represented.

The concept of granulation as a part of human cognition was first proposed and discussed by Zadeh (1997). Here, granulation involves partitioning of an object into granules, with a granule being a clump of elements drawn together by indistinguishability, equivalence, similarity or functionality.

Granulation can also be viewed as 'natural clustering', i.e., replacing a fine-grained universe by a coarse-grained one. This is a process like self-organization, self-production, Darwinian evolution, morphogenesis, group behavior-that are abstracted from natural phenomena. It is inherited in human thinking and rejoining process, and plays an essential role in human cognition.

1.3.2 *Information Granules*

Granular computing is regarded as a label of theories, methodologies, techniques and tools that make use of granules in the process of problem solving. It is a mode of computation in which the objects of computation are granular variables, not the data. There are two basic issues in granular computing, namely, construction of granules and computation with granules. According to Zadeh (1997), construction of a granule is based on the concept of generalized constraint. Computation or equivalently deduction is viewed as a sequence of operations involving combination, projection, qualification, propagation and counter-propagation of those constraints. There, the relationship between the granules was represented in terms of fuzzy graphs. If the elements within a granule are considered as a whole rather than individually, then some loss of information due to this granulation may occur. According to Pawlak (1998), the lost subsets of the universe can be approximated by applying the theory of rough set. An arbitrary set of a universe may not be a union of the same equivalence classes, but their lower and upper approximations give all the equivalence granules which are either a subset of the set or have nonempty intersection with the set, based on these, data analysis can be performed. Apart from fuzzy sets and rough sets, power algebra, interval number algebra and interval set algebra may be used for granular computing. Several types of operations on granules or computation procedures with granules are described in Yao (2000). The possible effectiveness of granular computing for handling big data has been discussed recently by Pal *et al.* (2015). Understanding of an image depends on its proper partitioning. Granulation can be useful to find segments/regions in an image. Depending on the application, granulation could be of different types with granules of equal or unequal size, although unequal granules are more natural for real life problems.

In this book, the tasks of video computing (e.g., tracking, conceptualization and object recognition) are dealt with granular computing where both formation of granules and computation with them under different situations are addressed in rough set theoretic framework. In the next section, we provide briefly the definition of rough set, its characteristics, and its relevance to image/video processing for handling uncertainty. This is followed by the basics of deep learning, internet of things (IoT) and the scope of the book.

1.4 Rough Sets

Theory of rough sets, as explained by Pawlak (1992), has become a popular mathematical framework for granular computing. The focus of the theory is on the ambiguity caused by limited discernibility of objects in the domain of discourse. Its key concepts are those of object 'indiscernibility' and 'set approximation'. Two major characteristics of the theory that have drawn the attention of applied researchers are uncertainty handling (using lower and upper approximations) and granular computing (using information granules). These characteristics made the theory useful in several areas of pattern recognition and machine learning, e.g., feature reduction and selection (Swirniaski, 2001; Komorouski *et al.*, 1999), image processing (Sen and Pal, 2009a, 2009b; Pal *et al.*, 2005), data mining and knowledge discovery (Komorouski *et al.*, 1999; Pedrycz and Song, 2014; Albanese *et al.*, 2014). Some basic definitions and features, as used in the work presented here, concerning this theory are given as follows.

1.4.1 *Definitions*

Suppose we have an information system $S = (U, A)$, where U is the universe and A is the set of attributes. For any set $B \subseteq A$, there is an equivalance relation $IND(B)$ such that $IND(B) = \{(x, y) \in U^2 | \forall p \in B, p(x) = p(y)\}$, where $p(x)$ function returns the value of the attribute p for data point x. The relation $IND(B)$ is called B-indiscernibility relation and any two points $(x, y) \in IND(B)$, i.e., satisfying the B-indiscernibility relation, indicate that x and y cannot be distinguishable using the attribute set B. Let the equivalence class of B-indiscernibility relation be denoted by $[x]_B$, and let $U|B$ denote all such equivalence classes. Here $[x]_B$ is called a 'granule' around the data point x, created by B-indiscernibility relation. (As stated before, a granule is a clump of objects which cannot be discriminated with a given attribute set.) Let us denote this granulated information system with $S_B = (U, A, [x]_B)$.

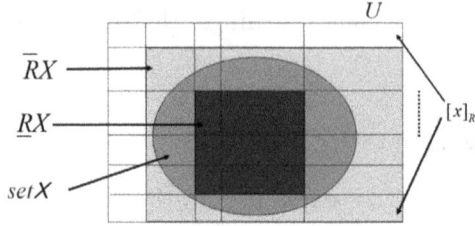

Fig. 1.2: Rough representation of a set X.

Let X be a set in the universe U ($X \subseteq U$) to be approximated based on the equivalence classes $[x]_B$ (i.e., granules) defined over B. Then, X can be approximated in terms of granules from inner and outer sides as *B-lower approximation* $\underline{B}X$ and *B-upper approximation* $\overline{B}X$, respectively. They are defined as follows:

$$\underline{B}X = \{x \in U : [x]_B \subseteq X\} \tag{1.1}$$

$$\overline{B}X = \{x \in U : [x]_B \cap X \neq \emptyset\} \tag{1.2}$$

$\underline{B}X$ represents the granules definitely belonging to X, while $\overline{B}X$ means granules definitely and possibly belonging to X. That means all the elements in $\underline{B}X$ can be certainly classified as members of X on the basis of the knowledge in B, while some objects in $\overline{B}X$ can only be classified as possible members of X on the basis of B.

The pictorial representation of a rough set is shown in Figure 1.2. Here, the oval shaped area represents the set X to be approximated over the small rectangles or the set of equivalent class $[x]_B$. The upper approximation $\overline{B}X$ of the set is represented by the larger light gray rectangular region and the lower approximation $\underline{B}X$ is represented by the dark gray and smaller rectangular region.

The roughness of the set X with respect to B (and $I_B = IND(B)$, the induced equivalence relation) can be characterized numerically (Pawlak, 1992) as:

$$R_B(X) = 1 - \frac{|\underline{B}X|}{|\overline{B}X|} \tag{1.3}$$

Equation (1.3) provides the measure of exactness of X. This means if $\underline{B}X = \overline{B}X$, roughness of the set X is 0, and X is exact with respect to B. If $R_B(X) > 0$, then X is roughly definable (i.e., X is vague with respect to B). This vague/inexact definition of X in the universe U, in terms of lower and upper approximations, signifies the incompleteness of knowledge about U.

1.4.2 *Reduct and Core*

It can be observed from the previous section that if the rough set is defined over a set of equivalence relations (B), it is a subset of the family of equivalent relations (A). The selection of the subset of equivalence relation plays a crucial role to approximate a set. It is true that all the attributes in the family of equivalence relation may not be necessary to define a proper set. Certain attributes in an information system may be redundant and can be eliminated without losing the essential classification/discrimination information. The procedure of eliminating those redundant equivalence relations from the point of view of rough sets is as follows.

Let A be the family of equivalent relations and let $b \in A$. b will be said *to be a dispensable* feature in A if $IND(A) = IND(A - \{b\})$, otherwise b is *indispensable* in A. The family of A is *independent* if each $b \in A$ is indispensable in **R**, otherwise it is dependent. Suppose, $B \subseteq A$, B is independent and $IND(B) = IND(A)$; then, B is called a *reduct* of A. A may have more than one reduct. The set of all indispensable relations in A is called the *core* of A. It can be shown that $CORE(A) = \bigcap RED(A)$. This is how the knowledge can be reduced from a knowledge base (Pawlak, 1992).

1.4.3 *Partial Dependency in Knowledge Base*

Theorizing based on drawing inference about the world is another (other than classification) important task. In other words, the problem is—how can knowledge be induced from a given knowledge base?

If there is not enough information available for classifying all the data points in a given set, then some part of that set can be classified by employing the knowledge of some other classifications. That part of the set is known as positive region of the classification with respect to the other classification. It can be expressed as:

$$POS_P(Q) = \bigcup_{X \in U/Q} \underline{P}X \tag{1.4}$$

where, P and Q are equivalence relations over U and Eq. (1.4) denotes P positive regions of Q. The dependency between P and Q is measured according to:

$$k = \gamma_P(Q) = \frac{|POS_P(Q)|}{|U|} \qquad (1.5)$$

$|.|$ denotes the cardinality of the sets in Eq. 1.5. If $k = 1$, then all elements in the universe can be classified into the elementary categories of U/Q by employing the knowledge of P. If $k \neq 1$, then only the elements within $POS_P(Q)$ can be classified.

1.4.4 *Indiscernibility Matrix*

Given an information table S, its indiscernibility matrix $M = M(x, y)$ is a $|U| \times |U|$ matrix, in which the element $M(x, y)$ for an object pair (x, y) is defined by:

$$M(x, y) = \{a \in A | I_a(x) \neq I_a(y)\} \qquad (1.6)$$

In Eq. (1.6), I_a is an information function that maps an object of U (x and y) to exactly one value. The indiscernibility matrix $(M(x, y))$ physically means that a pair of objects (x, y) can be distinguished (discerned) by any attribute of that matrix. That is, the pair (x, y) can be discerned if $M(x, y) \neq \emptyset$.

1.4.5 *Neighborhood Rough Sets*

Overlapping in class, region or concept is a usual characteristic in real life data. The neighborhood rough set (NRS) (Hu *et al.*, 2008; Du *et al.*, 2011), a new variant of Pawlak's rough set (Pawlak, 1992) (PaRS), deals with the concept of overlapping granules in numerical feature space. Formation of overlapping neighborhood granules around two points is shown in Figure 1.3, as an example.

The neighborhood of a point $x_i \in U$, in universe U may be represented as

$$\aleph(x_i) = \{x_j \in U : \triangle(x_i, x_j) \leq \delta\} \qquad (1.7)$$

where $\triangle$ is a distance function and δ is a threshold used for generating the neighborhood granules. Depending on the data, their values may remain the same or vary. The granules formed using Eq. 1.7 are

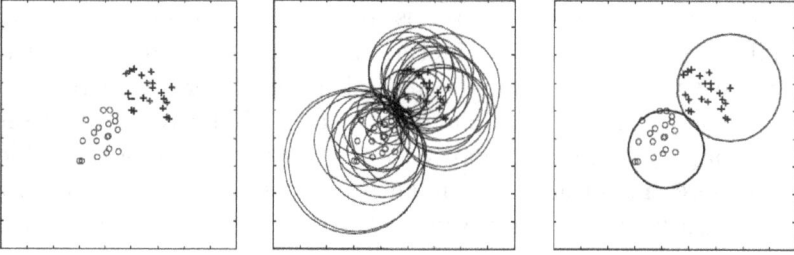

Fig. 1.3:   Example of formation of two neighborhood granules.

overlapping. That is, unlike PaRS a data point may belong to more than one granule in NRS. In other words, the constituting points of a granule may be shared by other granules, too.

Unlike PaRS, NRS can be characterized by two kinds of approximations of a set X. These are point-based $B_1(X)$ (type 1) and granule-based $B_2(X)$ (type 2), defined as

$$\underline{B_1}X = \{x \in U : \aleph(x) \subseteq X\} \qquad (1.8a)$$

$$\overline{B_1}X = \{x \in U : \aleph(x) \cap X \neq \emptyset\} \qquad (1.8b)$$

$$\underline{B_2}X = \{\aleph(x) \in U : \aleph(x) \subseteq X\} \qquad (1.8c)$$

$$\overline{B_2}X = \{\aleph(x) \in U : \aleph(x) \cap X \neq \emptyset\} \qquad (1.8d)$$

Accordingly, type 1 roughness R_1 is defined in terms of belonging of points (x), whereas type 2 roughness R_2 is defined in terms of belonging of the granules formed around x. If $\underline{R_1}X = \underline{R_2}X$ and $\overline{R_1}X = \overline{R_2}X$, then there is no overlapping neighborhood granule, i.e., granules are crisp, and NRS boils down to PaRS.

So far in Sections. 1.4.1 to 1.4.5, we have discussed the underlying concepts of rough set theory in terms of granules, some of its characteristic features, and roughness measure. Relevance of this theory to image processing and modeling uncertainty therein will be discussed in the following section.

1.4.6   *Rough Sets and Image Processing*

Here we first explain the relevance of rough sets and granular computing in image processing. Then several attempts made towards rough set theoretic image processing tasks are discussed.

In a gray image, nearby pixels and gray levels have limited discernibility. For example, a pixel in an image always tends to attain its intensity value as close to those of its neighbors. This results in the notion of rough resemblance of pixels. Similarly, rough resemblance of gray levels (i.e., a gray value is close to its adjacent values) exists there. This rough resemblance justifies the formation of granules within pixels and gray values, and hence the relevance of using granular computing in rough set framework, for image processing tasks.

Again, one may note that while the sinusoidal variation of gray values in an image results in fuzziness in boundaries, shape, etc. in object regions, the concept of roughness comes into picture because of the discretization (granularization) of the image space. The discretization makes an image be represented as a 2-D array of pixels (picture elements), where a pixel is characterized by its location in the image along with its intensity value.

The aforesaid fuzziness and roughness make the boundaries between object regions in a gray image often ill-defined and the definition of the object inexact; thereby resulting in both grayness and spatial ambiguities Pal *et al.* (2005). The uncertainty arising out of this can be handled by describing different objects as rough sets with upper (or outer) and lower (or inner) approximations. Here the concepts of upper and lower approximations can be viewed, respectively, as outer and inner approximations of an image region in terms of granules.

Accordingly, Pal *et al.* (2005) defined a rough image as a collection of pixels along with the equivalence relation to partition the image into sets of pixels lying within each non-overlapping window formed over the image. This definition is proven in Section 1.4.7. With this definition, the roughness of various partitions of an image can be computed using image granules for windows of different sizes, or image granules of arbitrary size and shape. Its details are given in Chapters 2 and 4.

There are a number of techniques for understanding, representating and processing the images and their segments and features as rough sets. Here we discuss a few of them. Wójcik (1994) first came up with a technique for edge enhancement using the theory of rough sets. Pal *et al.* (2005) first showed how the concept of rough set theoretic granular computing can be applied in image processing by defining an image as a rough set. The work includes representation of object–background as rough sets, formulation of 'rough entropy', and image segmentation with rough entropy maximization. Mushrif and Ray (2008) proposed a method for color image segmentation using color histograms and histones. In rough set theoretic notion, the histogram was correlated with the lower approximation and the histon was correlated with upper approximation in the work. The roughness measure corresponding to every intensity level was then computed and a thresholding method was applied for image segmentation. Hassanien and Abraham (2008) illustrated how rough set can be successfully integrated with mathematical morphology and provided a more effective hybrid approach to resolve medical imaging problems. Here the rough set approach for attribute reduction and rule generation is used and rough morphology is designed for discrimination of different regions of interest to detect whether those regions represent the cancerous one. Several rough clustering algorithms (Lingras, 2007) have also been developed and applied in medical imaging problems. Sen and Pal (2009a) defined generalized rough-fuzzy sets and applied them for image segmentation. Here they formulated a class of entropy measures based on rough set theory and its certain generalizations, quantified grayness and spatial ambiguities in images, and defined an average image ambiguity measure. The generalization involves incorporation of fuzziness in defining the set and/or granules. In another work (Sen and Pal, 2009b), they proposed a method of rough-fuzzy histogram thresholding. Here they carried out the tasks of bi-level thresholding, multilevel thresholding, object/background separation and edge extraction over images. Meher and Pal (2011) formulated a new rough-wavelet granular space based model for land cover classification of multi-spectral remote sensing image where they formulated class-dependent granules in wavelet domain using shift-invariant wavelet transform. Although the theory of rough sets has been used in several areas of image processing, the merits of neighborhood rough sets have not been explored much. They have only been used recently for band selection in hyper-spectral images (Liu *et al.*, 2016b). Besides, investigation on using rough sets and granular computing in video processing is very scanty, although it has been done adequately in image processing for a long time.

1.4.7 *Rough Sets and Uncertainty Modeling*

It is seen from Eq. (1.3) how the inexactness/vagueness of a rough set can be quantified to signify the incompleteness of knowledge about the universe U. Here we explain the uncertainty modeling in an image based on this concept, considering the gray tone image as a rough set.

1.4.7.1 *Image as Rough Set*

Let the universe U be an image consisting of a collection of pixels. Then if we partition U into a collection of non-overlapping windows (of size $m \times n$, say), each window can be considered as a granule G. In other words, the induced equivalence classes $I_{m \times n}$ have $m \times n$ pixels in each non-overlapping window. Given this granulation, object regions in the image can be approximated by rough sets. Such an example image, where object/background are approximated as rough sets, is shown in Fig. 1.4.

Let us consider an object–background separation (a two class) problem of an $M \times N$, L level image. Let prop(B) and prop(O) represent two properties (say, gray level intervals $0, 1, \cdots, T$ and $T + 1, T + 2, \cdots, L - 1$) that characterize background and object

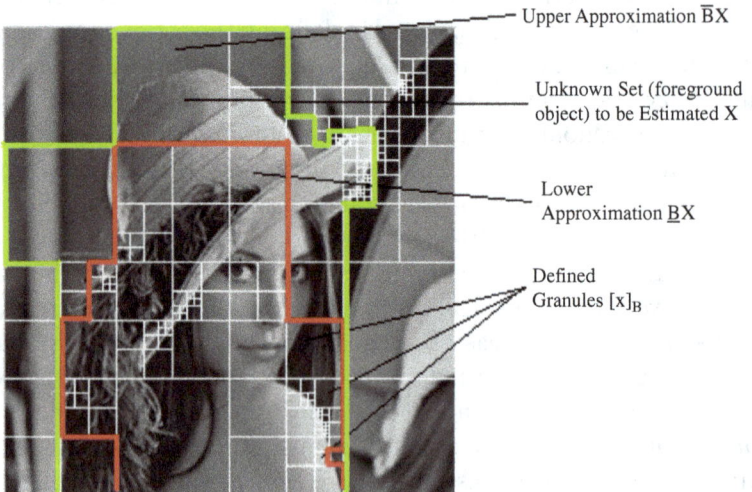

Upper Approximation $\overline{B}X$

Unknown Set (foreground object) to be Estimated X

Lower Approximation $\underline{B}X$

Defined Granules $[x]_B$

Fig. 1.4: Example of rough set representation of an image.

regions, respectively. Given this framework, object and background can be viewed as two sets with their rough representation as follows: The inner approximation of the object $(\underline{O}_T)$:

$$\underline{O}_T = \left\{ \bigcup_i G_i \mid P_j > T, \ \forall j = 1, \ldots mn, \text{ and} \right.$$

$$\left. P_j \text{ is a pixel belonging to } G_i \right\}$$

$$= \text{set of granules with all the pixels values } > T$$

Outer approximation of the object $(\overline{O}_T)$:

$$\overline{O}_T = \left\{ \bigcup_i G_i, \ \exists j, \ j = 1 \ldots mn \ s.t. \ P_j > T, \text{where} \right.$$

$$\left. P_j \text{ is a pixel in } G_i \right\}$$

$$= \text{set of granules with at least one pixel value } > T$$

Inner approximation of the background $(\underline{B}_T)$:

$$\underline{B}_T = \left\{ \bigcup_i G_i \mid P_j \leq T, \ \forall j = 1, \ldots mn, \text{ and} \right.$$

$$\left. P_j \text{ is a pixel belonging to } G_i \right\}$$

$$= \text{set of granules with all the pixels values } < T$$

Outer approximation of the background $(\overline{B}_T)$:

$$\overline{B}_T = \left\{ \bigcup_i G_i, \ \exists j, j = 1 \ldots mn \ s.t. \ P_j \leq T, \text{ where} \right.$$

$$\left. P_j \text{ is a pixel in } G_i \right\}$$

$$= \text{set of granules with at least one pixel value } < T.$$

Therefore, the rough set representation of the image (i.e., object O_T and background B_T) for a given $I_{m \times n}$ depends on the value of T.

Let the roughness of object O_T and background B_T be defined as

$$R_{O_T} = 1 - \frac{|\underline{O}_T|}{|\overline{O}_T|} = \frac{|\overline{O}_T| - |\underline{O}_T|}{|\overline{O}_T|}$$

$$R_{B_T} = 1 - \frac{|\underline{B}_T|}{|\overline{B}_T|} = \frac{|\overline{B}_T| - |\underline{B}_T|}{|\overline{B}_T|}$$

$$(1.9)$$

where $|\underline{O}_T|$ and $|\overline{O}_T|$ are the cardinality of the sets $\underline{O}_T$ and $\overline{O}_T$, and $|\underline{B}_T|$ and $|\overline{B}_T|$ are the cardinality of the sets $\underline{B}_T$ and $\overline{B}_T$, respectively.

1.4.8 *Rough Entropy*

Rough Entropy (RE) of an image can be defined as Pal *et al.* (2005)

$$RE_T = -\frac{e}{2}[R_{O_T} \, log_e(R_{O_T}) + R_{B_T} log_e(R_{B_T})].$$

$$(1.10)$$

Its plot for various values of R_{O_T} and R_{B_T} is shown in Fig. 1.5.

(1) The value of RE_T lies between 0 and 1.
(2) RE_T has a maximum value of unity when $R_{O_T} = R_{B_T} = 1/e$, and minimum value of zero when $R_{O_T}, R_{B_T} \in \{0, 1\}$.

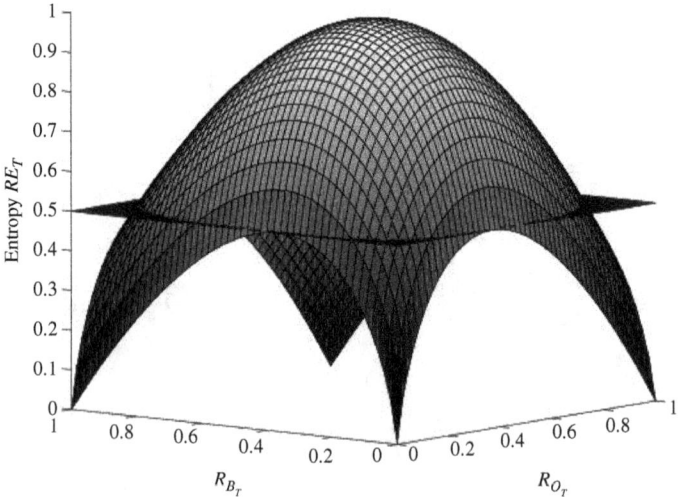

Fig. 1.5: Plot of rough entropy for various values of roughness of the object and background.

(3) (a) Since the boundary pixels are common for both object and background, we have $\overline{O}_T - \underline{O}_T = \overline{B}_T - \underline{B}_T = Q_T$, say. Therefore,

$$R_{O_T} = R_{B_T}, \quad iff \ |\underline{O}_T| = |\underline{B}_T|.$$

Under this condition, the distribution of RE_T on the diagonal (joining $(0,0)$ and $(1,1)$) is shown in Fig. 1.6, where RE_T attains a maximum value of unity at $R_{O_T} = R_{B_T} = 1/e$.

(b) When $|\underline{O}_T| < |\underline{B}_T|$, then $R_{O_T} > R_{B_T}$ and when $|\underline{O}_T| > |\underline{B}_T|$, then $R_{O_T} < R_{B_T}$.

In either case, RE_T will decrease from its maximum value of unity and will reach a value of zero at $(0,0), (0,1), (1,0)$ and $(1,1)$ in the (R_{O_T}, R_{B_T}) plane (Fig. 1.5).

Equation (1.10), defining rough image entropy, quantifies the incompleteness of knowledge about the image with respect to the definability of the object–background region. This entropy measure is defined by Pal *et al.* (2005). One can modify the definition according to the need and purpose keeping the underlying concept same. One such modification is suggested by Sen and Pal (2009a). Another modification may be made, as in Eq. (1.11), by choosing different values of BASE in Eq. (1.10). This modification provides flexibility to choose the right value of the *BASE* for the image under consideration with the amount of noise present in the image. By trial and error it was found (Chakraborty *et al.*, 2013) that for most of the images a *BASE* value of '10' is quite suitable. However, one can choose the value 'e' or '2' as the *BASE*.

$$RE_T = -\frac{BASE}{2}\left[R_{O_T} \, log_{(BASE)}\left(R_{O_T}\right) \right.$$

$$\left. + R_{B_T} log_{(BASE)}\left(R_{B_T}\right) \right] \tag{1.11}$$

Further, in Eq. (1.11), the maximum value of '1' will be attained at $1/(BASE)$, which is a function of $BASE$, providing necessary flexibility to consider the noise, while computing the rough entropy.

Besides using granular computing techniques, we have described in this book the development and merits of some hybrid tools involving granular computing. Deep learning is one such tool that we used here to solve a few tasks. The concept of deep learning and its impact on image processing are discussed in brief in the following section.

1.5 Deep Learning

For the past decade, deep learning (in-depth learning) has emerged as a new area of machine learning. A machine learning method with the following characteristics is called 'deep learning': (i) using a cascade of multiple layers of nonlinear processing units for feature extraction and transformation where the output of each layer is the input to its next layer, (ii) learning is supervised and/or unsupervised, and (iii) learning multiple levels of representations that correspond to different layers from a hierarchy of concepts of abstraction or multi-dimensional learning in other words.

A neural network (NN) architecture is found to be most suitable with the aforesaid features. It is made up of several neurons. Input neurons get triggered from the environment while the other neurons get triggered through weighted links from previously active neurons. In order to extract the complex representation from rich sensory

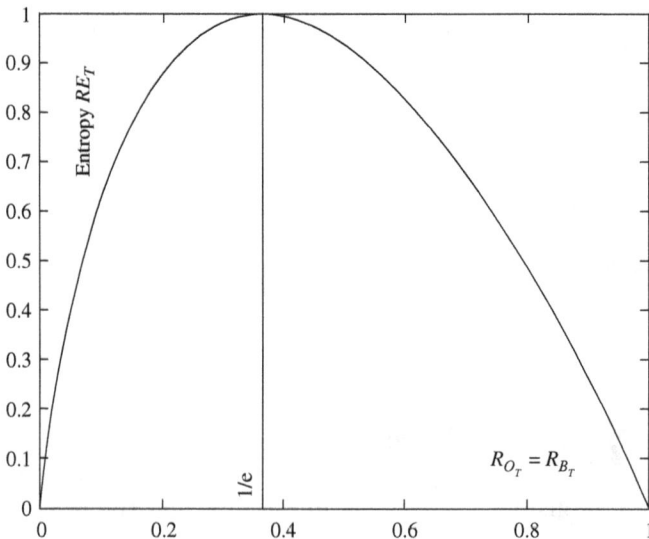

Fig. 1.6: Plot of rough entropy for the values (0, 0) to (1, 1) on the diagonal of Fig. 1.5 (i.e., when $R_{O_T} = R_{B_T}$).

inputs, human information processing mechanisms suggest the need for deep architectures (LeCun *et al.*, 2015).

1.5.1 *Deep Learning and Image Processing*

In deep learning architecture, the problem of object recognition can be regarded as a task of labeling different objects in an image with the correct class as well as predicting the bounding boxes along with probability. Different structures of deep neural networks on this problem have been proposed (Erhan *et al.*, 2014; Ren *et al.*, 2015; Redmon *et al.*, 2016; Liu *et al.*, 2016a). A convolutional neural network (CNN, or ConvNet) (Krizhevsky *et al.*, 2012) is one of the most popular algorithms for deep learning with images and video. Like other neural networks, a CNN is composed of an input layer, an output layer, and several hidden layers in between. These layers perform one of the three types of operations on the data, i.e., convolution, pooling or rectified linear unit (ReLU). Convolution puts the input images through a set of convolutional filters (Kavukcuoglu *et al.*, 2010), each of which activates certain features from the images. Pooling (Krizhevsky *et al.*, 2012) simplifies the output by performing nonlinear downsampling, reducing the number of parameters that the network needs to learn about. Rectified linear unit (ReLU) (LeCun *et al.*, 2015) allows for faster and more effective training by mapping negative values to zero and maintaining positive values. These three operations are repeated over tens or hundreds of layers, with each layer learning to detect different features. CNNs have been used for motion detection and object recognition (Karpathy *et al.*, 2014; Ji *et al.*, 2013; Gan *et al.*, 2018; Held *et al.*, 2016). Deep learning has dramatically improved the state-of-the-art in object recognition (LeCun *et al.*, 2015). However, deep learning often requires hundreds or thousands of images for the best results unlike the conventional (Shallow) learning. Therefore, deep learning is computationally intensive and requires a high-performance graphical processing unit (GPU).

1.5.2 *Deep Learning Models*

There are different types of deep learning models developed so far (Schmidhuber, 2015; Cios, 2018; Karpathy, 2015). The most popular architectures that are used to form a deep network are: convolutional neural network (CNN), recurrent neural network (RNN) and recursive neural network. The basic differences among these three architectures are: (i) CNN is basically a standard neural network. It is extended with weight sharing in different nodes on the space. CNN architecture is applied to recognize images by having convolutions inside. (ii) An RNN is also a standard neural network. But here the feed is carried out into the next time step instead of next spatial layer of the same time step. This RNN is mainly applied on time-variant signals, that is the data which varies with time. A speech signal, text or video signal are the examples of time-variant signals. The network has a short memory to keep the information till next time step. (iii) A Recursive Neural Network is more like a hierarchical network where there is really no time aspect to the input sequence, but the input has to be processed hierarchically in a tree fashion. The most effective learning and updation methods developed so far and applied in all the aforementioned architectures are described as follows:

- Back Propagation: This is one of the most popular deep learning methods. Here the optimization is carried out by back propagating the error gradient in each iteration.
- Learning Rate Decay: Here the learning rate of the network gets reduced with time. There are two ways of decreasing the rate: (i) gradually with each epoch and (ii) suddenly with punctuated epoch.
- Drop Out: It is used to avoid overfitting and to have computational gain in the huge network. Random nodes in each layer with their connections are dropped out from further computation in this method. This results in a huge computational gain in testing.
- Max Pooling: The input data are downsampled and represented with reduced dimensionality in this method. A max filter is applied to the non-overlapping regions of input data in this technique. Computational gain is achieved by carrying out the computation with lower dimension data.
- Transfer Learning: This method is very effective for vision-based problems. It is mostly applicable with CNN. Initially, each layer of a CNN builds higher representation of features. That is, the earlier levels of the CNN consume such features common to different classes, whereas the layer closer to the last layer consumes the features specific to some certain class. Once the last layer is ready to

recognize a specific class, the last layer(s) is chopped off from the network and a different dataset is fed to the network to train the network for some other class of objects. This way, a single network can be used for different classes of object recognition and results in a computational gain.

While the effectiveness of most of the algorithms of video processing, described in this book is validated with some existing benchmark datasets, a few of them are also tested with real-time data acquisition. Real-time data acquisition is done particularly when testing the algorithms in an architecture of internet of things (IoT). A brief overview of IoT is provided in the next section to make the readers familiar with the concept.

1.6 Internet of Things (IoT)

The internet of things (IoT) is a network of physical devices that are embedded with sensors, software and electronics. IoT refers to the networked interconnection of everyday objects, which are often equipped with intelligence (Greengard, 2015; Xia *et al.*, 2012; Thiesse *et al.*, 2009; Ngu *et al.*, 2017).

1.6.1 *IoT and Video Processing*

The term internet of things (IoT) was first introduced by Astron [Gershenfeld (1999)] in 1999. There he said, 'If we had computers that knew everything there was to know about things – using data they gathered without any help from us – we would be able to track and count everything, and greatly reduce waste, loss and cost. We would know when things needed replacing, repairing or recalling, and whether they were fresh or past their best. We need to empower computers with their own means of gathering information, so they can see, hear and smell the world for themselves, in all its random glory.' This is precisely what IoT platforms do for us. They enable devices/objects to observe, identify and understand a situation or the surroundings without being dependent on human help. This enables these things to connect, collect and exchange data.

The IoT extends internet connectivity beyond traditional devices like desktop and laptop computers, smartphones and tablets to a diverse range of devices and everyday things that utilize embedded technology to communicate and interact with the external environment, all via the internet. A thing in the IoT can be a person with a heart monitor implant, a farm animal with a biochip transponder, an automobile that has built-in sensors to alert the driver when tire pressure is low or any other natural or man-made object that can be assigned an IP address and is able to transfer data over a network. An IoT ecosystem consists of web-enabled smart devices that use embedded processors, sensors and communication hardware to collect, send and act on data they acquire from their environments. IoT devices share the sensor data they collect by connecting to an IoT gateway or other edge device where data is either sent to the cloud to be analyzed or analyzed locally. Sometimes, these devices communicate with other related devices and act on the information they get from one another. The devices do most of the work without human intervention, although people can interact with the devices — for instance, to set them up, give them instructions or access the data.

1.6.2 *IoT: Features and Benefits*

The main features of IoT are as follows:

- AI: It is one of the key features of IoT to make anything 'smart'. That is, it enhances every aspect of life with the power of data collection and processing in the networks with the ability of intelligent decision-making. Here comes the requirement of advanced artificial intelligence algorithms.
- Connectivity: Building up proper connections among sensor, processor and user is another goal in IoT. Besides, the network should be smaller and cheaper, too. IoT creates these small networks between its system devices.
- Sensors: IoT acquires information from the environment through its sensors. These are the instruments which transform IoT from a standard passive network of devices into an active system capable of real-world integration.
- Small Devices: A smart network requires smaller, cheaper and powerful devices. Some purpose-built small devices are developed to deliver the precision, scalability and versatility of IoT.

IoT encourages companies to rethink the ways they approach their businesses, industries and markets and gives them the tools to improve their business strategies. The internet of things aims to provide a number of benefits to the users. These include:

- monitoring of overall business processes;
- improving the customer experience;
- saving time and money;
- enhancing employee productivity;
- integrating and adapting intelligent business models;
- making better business decisions.

1.6.3 *IoT Applications*

There are numerous real-world applications of the IoT. It has a huge range starting from consumer IoT and enterprise IoT to manufacturing and industrial IoT (IIoT). The primary applications of IoT include the following:

- Smart homes: these are equipped with smart thermostats, smart appliances and connected heating, lighting and electronic devices. All the devices can be controlled remotely via computers, smartphones or other mobile devices.
- Wearables: These are the devices with sensors and software that can collect and analyze user data, send messages to other technologies about the users with the aim of making users' lives easier and more comfortable.
- Healthcare: IoT offers many benefits in healthcare. This includes the ability to monitor patients more closely with the data that gets generated from the sensors attached with the patients and analyze it. Hospitals often use IoT systems for the surveillance of the patients.
- Smart buildings: It can reduce energy costs using sensors. For example, with detection of the total occupants in a room the system can adjust room temperature automatically and turn on or off the air conditioner.
- Agriculture: IoT-based smart farming systems can help to monitor light, temperature, humidity and soil moisture of crop fields using connected sensors.

- Smart city: IoT sensors and decision-making systems help to build a smart city. A smart city includes smart street lights, smart traffic control systems, energy conservation, monitoring and addressing environmental concerns, and improving sanitation.

It is clear from the aforementioned applications of IoT that some IoT frameworks seem to focus on acquisition of real-time data and make them interact with some other 'things' in order to have a real-time intelligent network. The usage and application of IoT is increasing day by day based on different requirements of the users. In Chapter 7 of the book, we have presented some results of investigations demonstrating the effectiveness of our video conceptualization algorithm in IoT with real-time video acquisition and decision-making with physical devices.

In this chapter we have discussed, so far, different concepts, features, challenges and theories related to various tasks of video processing. These will be elaborated and probable solutions will be formulated in different chapters of this book. The next section provides an overview (scope) and organization of the rest of the book.

1.7 Scope and Organization of the Book

As discussed in Section 1.2, there are several unsolved issues in the task of video analysis. These include handling uncertainties arising from different types of ambiguities in video sequences that could not be efficiently done till now. The present book provides some new results, both theoretical and experimental, of investigation describing various techniques and methodologies to handle some of those issues. It deals mainly with the problems arising in: (i) tracking due to the absence of complete information regarding the number of moving object(s), background, newly appeared object(s), moving-to-static object(s), partly moving object(s), partially occluded or overlapped object(s), and fully occluded or overlapped moving object(s); (ii) object recognition due to high computation time; and (iii) video understanding due to the absence of large amount of labeled dataset, and huge computation time. All these issues are addressed in different granular computing frameworks along with the theory of rough sets and deep learning. Besides, experimentation on testing the effectiveness of the video conceptualization method with real-time video acquisition in an IoT architecture is described. Development of several new quantitative indices is explained to evaluate the performance of tracking, object recognition and scene description. Tracking indices do not require any ground truth information unlike most of the existing indices.

Details of the methodologies, their features, experimental results performed on different types of video sequences sensed by RGB, Kinect and IR sensors, and the associated novelties are summarized in the following under different chapter headings.

A spatio-temporal method for object tracking in rough set theoretic granular computing framework is described in Chapter 2, among others. The method involves initial labeling of objects, and therefore it is a partially supervised approach. Here the concept of lower and upper approximations of rough set theory is used to form spatio-color granules. Spatial segmentation is performed using rough entropy maximization, where the quad-tree decomposition is used to result in unequal image granulation, which is closer to natural granulation. The method of temporal segmentation involves background estimation and subtraction by defining a three-point estimation based on Beta Distribution for background estimation. Reconstruction and tracking of the object in the target frame is performed by estimating the color distribution of the target after combining the two (spatial and temporal) segmentation outputs. These methods for spatial and temporal segmentation are seen to be superior to several methods of a similar category. The accuracy of reconstruction has been significantly high. However, this method works well if the moving object and background have different gray level distributions, as its performance is dependent on the results of spatial segmentation of image frames.

Chapters 3 and 4 deal with the task of tracking multiple moving objects. The merits of neighborhood rough sets in granular computing are exploited there.

Chapter 3 explores the merits of rough rule-base in video tracking. The method described is unsupervised where no information is initially available. A new concept of neighborhood granule formation over images is introduced. These granules are of arbitrary shape and size, unlike other existing granulation techniques, and hence more natural. A way of forming spatio-temporal 3-D granules of arbitrary shape and size is explained. With these granules, a rough rule-base is formed and used for video tracking to deal with the uncertainties and incompleteness as well as to gain in computation time. In the second part of the chapter, an adaptive neighborhood granular rough rule-base is formed which proves to be effective in reducing the indiscernibility of the rule-base. In the process of development of adaptive rule-base, several concepts and operations related to it are described. The flow graph, which is used to develop adaptive rule-base, enables in defining an intelligent technique for rule-base adaptation. Two relevant features, namely, expected flow graph and mutual dependency between flow graphs are described here to make the flow graph applicable in the tasks of both training and validation. All these techniques are performed in neighborhood granular level. The rough flow graph-based adaptive granular rule-based system is capable of handling the uncertainties and incompleteness in frames, able to overcome the incompleteness in information that arises without initial manual interactions and in providing superior performance and gaining in computation time.

The problem of tracking overlapping/occluded objects in unsupervised mode is dealt with in Chapter 4 with the concepts of lower–upper approximations of neighborhood rough sets. It involves the design of a neighborhood rough filter, as used for initial labeling of continuous moving object(s) even in the presence of several variations in different feature spaces. Filter output provides an estimation of the locations and color models of the object(s) using their lower–upper approximations in spatio-color neighborhood granular space. Object location in the next frame is predicted through formation of velocity neighborhood granules and acceleration neighborhood granules over the filter output estimation. These granule also accelerated the tracking process. The uncertainties arising due to the occurrence of overlapping/occlusion in a video sequence are handled with a novel concept, namely, 'intuitionistic entropy' which has two components, viz., neighborhood rough entropy and neighborhood probabilistic entropy. This unsupervised method of tracking performs equally well even when compared with some of the state-of-the-art partially supervised methods, while showing superior performance during total occlusion.

The explainability/trustability of different tracking algorithms is studied in Chapter 5. There already exist several indices to quantify the reliability of a tracking algorithm in different scenarios. But most of them require ground truth information. The indices that work without the ground truth information need to have the estimated trajectory. Here we have described a few indices that work without ground truth information and do not estimate the trajectory either. Three of them are defined based on rough sets, feature similarity and Bhattacharya distance for videos with only RGB information to evaluate the performance of tracking and detect the mistracked frames. Two others are to evaluate the performance of tracking using the merits of depth-feature or D-feature and granulation. An experimental study of the trustability of different tracking algorithms under different environments is provided.

The tasks of object recognition from a scene and video motion analysis are considered in Chapters 6 and 7, respectively.

Chapter 6 deals with the problems of motion detection, object recognition and scene description using deep learning in the framework of granular computing and Z-numbers. Since deep learning is computationally intensive, whereas granular computing, on the other hand, leads to computation gain, a judicious integration of their merits makes the learning mechanism computationally efficient. Such a system based on convolutional neural network (CNN) is described here. Further, it is shown how the concept of Z-numbers can be used to quantify the abstraction of semantic information in interpreting a scene, where subjectivity is of major concern, through recognition of its constituting objects. The said system involves recognition of both static objects in the background and moving objects in the foreground separately. Rough lower and upper approximations are used in defining object and background models. During deep learning, instead of scanning the entire image pixel by pixel in the convolution layer, it scans only the representative pixel of each granule. This results in a significant gain in computation time. Arbitrary shaped and sized granules, as expected, perform better than regular shaped rectangular granules or fixed sized granules. The method of tracking is able to deal efficiently with various challenging cases, e.g., tracking partially overlapped objects and suddenly appearing objects. Overall, the granulated system shows a balanced trade-of between speed and accuracy as compared to pixel level learning in tracking and recognition. The concept of using Z-numbers, in providing a granulated linguistic description of a scene, is unique. This gives a more natural interpretation of object recognition in terms of certainty towards scene understanding.

Chapter 7 mainly focuses on a methodology for identifying the occurrence of some events in a video without any prior knowledge. This unsupervised application is named as 'conceptualization'. Its significance in internet of things (IoT) is demonstrated. The method is applicable for the video sequences that are acquired by simple static RGB sensors. Here the video sequences are initially granulated with 'motion granules', a new concept, and then modeled as rough sets over this granulation for object/background estimation. The decision-making on estimation is based on quantifying the rough approximation with a measure called motion entropy. Its value reflects the amount of uncertainty present in the motion of each individual moving object. Therefore, it enables the classification of random movements (noise), continuous or predictable movements, and sudden changes or un-predictable movements in order to conceptualize the content of the video grossly, and to identify which frames of the sequence carry most of the information to analyze them further. Results on the effectiveness of this model in classifying the movements and detecting occurrence of events are shown for both offline and real-time video sequences. As a real-life application, an IoT architecture with physical devices is described to test the significance of the algorithm in identifying the frames containing unusual movements.

Besides, some other relevant methods and algorithms, tools and definitions, as required to understand the concerned topic, are mentioned in each chapter for the convenience of the readers.

Bibliography

Albanese, A., Pal, S. K. and Petrosino, A. (2014). Rough set, kernel set and spatio-temporal outlier detection, *IEEE Trans. Knowl. Data Engg.* **26**(1): 194–207.

Ali, A. and Aggarwal, J. K. (2001). Segmentation and recognition of continuous human activity, *IEEE Workshop on Detection and Recognition of Events in Video*, pp. 28–36.

Bai, T., Li, Y.-F. and Zhou, X. (2015). Learning local appearances with sparse representation for robust and fast visual tracking, *IEEE Trans. Cyberns.* **45**(4): 663–675.

Black, J., Ellis, T. and Rosin, P. (2003). A novel method for video tracking performance evaluation, *Proc. Joint IEEE Int. Workshop on VS-PETS*, pp. 125–132.

Borges, P., Conci, N. and Cavallaro, A. (2013). Video-based human behavior understanding: A survey, *IEEE Trans. CSVT* **23**(11): 1993–2008.

Brand, M. (1996). Understanding manipulation in video, *Proc. Second Intl. Conf. AFGR*, IEEE, Killington, VT, pp. 94–99.

Briechle, K. and Hanebeck, U. D. (2001). Template matching using fast normalized cross correlation, *Proc. SPIE* **4387**: 95–102.

Burgos-Artizzu, X., Dollar, P., Lin, D. and Anderson, D. (2012). Social behavior recognition in continuous video, *IEEE CVPR*, Providence, RI, pp. 1322–1329.

Cai, M., Lu, F. and Gao, Y. (2018). Desktop action recognition from first-person point-of-view, *IEEE Trans. Cyberns.* 1–13. DOI: 10.1109/TCYB.2018.2806381.

Chakraborty, D., Shankar, B. U. and Pal, S. K. (2013). Granulation, rough entropy and spatiotemporal moving object detection, *Appl. Soft Comput.* **13**(9): 4001–4009.

Cheung, S. S. and Kamath, C. (2004). A robust techniques for background subtraction in urban traffic video, *Proc. Video Communications and Image Processing, ISPIE Electronic Imaging* **5308**: 881–892.

Chi, Z., Li, H., Lu, H. and Yang, M.-H. (2017). Dual deep network for visual tracking, *IEEE Tran. Image Proc.* **26**(4): 2005–2015.

Cios, K. J. (2018). Deep neural networks—A brief history, *In Advances in Data Analysis with Computational Intelligence Methods—Dedicated to Professor Jacek Zurada* pp. 183–200.

Comaniciu, D., Ramesh, V. and Meer, P. (2003). Kernel-based object tracking, *IEEE Trans. PAMI* **25**(5): 564–575.

Cucchiara, R., Grana, C., Piccardi, M. and Prati, A. (2003). Detecting moving objects, ghosts and shadows in video streams, *IEEE Trans. PAMI* **25**: 1337–1342.

Čehovin L., Leonardis, A. and Kristan, M. (2016). Visual object tracking performance measures revisited, *IEEE Trans. Image Proc.* **25**: 1261–1274.

Davis, J. and Sharma, V. (2007). Background-subtraction using contour-based fusion of thermal and visible imagery, *Comput. Vis. Image Underst.* **106**: 162–182.

Dey, B. and Kundu, M. K. (2013). Robust background subtraction for network surveillance in h.264 streaming video, *IEEE Trans. Circ. Sys. Vid. Tech.* **23**(10): 1695–1703.

Du, Y., Hu, Q., Zhu, P. and Ma, P. (2011). Rule learning for classification based on neighborhood covering reduction, *Inf. Sci.* **181**(24): 5457–5467.

Erhan, D., Szegedy, C., Toshev, A. and Anguelov, D. (2014). Scalable object detection using deep neural networks, *Proceedings of the IEEE Conference on Computer Vision and Pattern Recognition*, pp. 2147–2154.

Fang, H., Jiang, J. and Feng, Y. (2006). A fuzzy logic approach for detection of video shot boundaries, *Pattern Recognit.* **39**: 2092–2100.

Fang, Y., Yuan, Y., Li, L., Wu, J., Lin, W. and Li, Z. (2017). Performance evaluation of visual tracking algorithms on video sequences with quality degradation, *IEEE Access* **5**: 2430–2441.

Felzenszwalb, P. F., Girshick, R. B., McAllester, D. and Ramanan, D. (2010). Object detection with discriminatively trained part-based models, *IEEE Trans. PAMI* **32**(9): 1627–1645.

Fergus, R., Perona, P. and Zisserman, A. (2003). Object class recognition by unsupervised scale-invariant learning, *IEEE Conference on Computer Vision and Pattern Recognition (CVPR)*, IEEE.

Gan, W., Lee, M.-S., Wu, C.-H. and Kuo, C.-C. J. (2018). Online object tracking via motion-guided convolutional neural network (mgnet), *J. Vis. Commun. Image Represent.* **53**: 180–191.

Gershenfeld, N. (1999). *When Things Start to Think*, Henry Holt and Co., Inc., New York, NY, USA.

Greengard, S. (2015). *The Internet of Things*, MIT Press.

Gupta, A., Srinivasan, P., Shi, J. and Davis, L. (2009). Unsupervised video understanding by reconciliation of posture similarities, *IEEE CVPR*, Miami, FL, pp. 2012–2019.

Hassanien, A. E. and Abraham, A. (2008). Rough morphology hybrid approach for mammography image classification and prediction, *Int. J. Comput. Intell. Appl.* **7**(1): 17–42.

Hassanien, A. E., Abraham, A., Peters, J. F. and Schaefer, G. (2008). Overview of rough-hybrid approaches in image processing, *Proc. IEEE Conf. on Fuzzy Systems*, IEEE Press, N. J., pp. 2135–2142.

Held, D., Thrun, S. and Savarese, S. (2016). Learning to track at 100 fps with deep regression networks, *European Conference on Computer Vision*, Springer, pp. 749–765.

Henriques, J. F., Caseiro, R., Martins, P. and Batista, J. (2015). High-speed tracking with kernelized correlation filters, *IEEE Trans. PAMI* **37**(3): 583–596.

Hu, Q., Yu, D., Liu, J. and Wu, C. (2008). Neighborhood rough set based heterogeneous feature subset selection, *Inf. Sci.* **178**(18): 3577–3594.

Huang, C.-M. and Fu, L.-C. (2011). Multitarget visual tracking based effective surveillance with cooperation of multiple active cameras, *IEEE Trans. Syst. Man Cybern. Part B* **41**(1): 234–247.

Janoch, A., Karayev, S., Jia, Y., Barron, J. T., Fritz, M., Saenko, K. and Darrell, T. (2011). A category-level 3-d object dataset: Putting the kinect to work, *ICCV Workshop on Consumer Depth Cameras for Computer Vision*.

Ji, S., Xu, W., Yang, M. and Yu, K. (2013). 3D convolutional neural networks for human action recognition, *IEEE Trans. PAMI* **35**(1): 221–231.

Karpathy, A. (2015). The Unreasonable Effectiveness of Recurrent Neural Networks. http://karpathy.github.io/2015/05/21/rnn-effectiveness/.

Karpathy, A., Toderici, G., Shetty, S., Leung, T., Sukthankar, R. and Fei-Fei, L. (2014). Large-scale video classification with convolutional neural networks, *Proceedings of the IEEE conference on Computer Vision and Pattern Recognition*, pp. 1725–1732.

Kasturi, R., Goldgof, D., Soundararajan, P., Manohar, V., Garofolo, J. and Bowers, R. (2009). Framework for performance evaluation of face, text, and vehicle detection and tracking in video: Data, metrics, and protocol, *IEEE Trans. PAMI* **31**(2): 319–336.

Kavukcuoglu, K., Sermanet, P., Boureau, Y.-L., Gregor, K., Mathieu, M. and Cun, Y. L. (2010). Learning convolutional feature hierarchies for visual recognition, *Adv. Neural Inf. Process. Syst.* pp. 1090–1098.

Komorouski, J., Pawlak, Z., Polkowski, L. and Skowron, A. (1999). Rough sets: a tutorial, *in* S. K. Pal and A. Skowron (eds.), *Rough Fuzzy Hybridization: A New Trend In Decision-Making*, Springer, Singapore, pp. 3–98.

Krizhevsky, A., Sutskever, I. and Hinton, G. E. (2012). Imagenet classification with deep convolutional neural networks, *Adv. Neural Inf. Process. Syst.* pp. 1097–1105.

Kwon, J. and Lee, K. M. (2014). Tracking by sampling and integrating multiple trackers, *IEEE Tran. PAMI* **36**(7): 1428–1441.

Lai, K., Bo, L., Ren, X. and Fox, D. (2011). A large-scale hierarchical multiview rgb-d object dataset, *ICRA*, pp. 1817–1824.

LeCun, Y., Bengio, Y. and Hinton, G. (2015). Deep learning, *Nature* **521**(7553): 436.

Leibe, B., Leonardis, A. and Schiele, B. (2008). Robust object detection with interleaved categorization and segmentation, *IJCV* **77**(1): 259–289.

Li, X., Shen, C., Dick, A. R., Zhang, Z. M. and Zhuang, Y. (2016). Online metric-weighted linear representations for robust visual tracking, *IEEE Trans. Pattern Anal. Mach. Intell.* **38**(5): 931–950.

Lingras, P. (2007). Trans. on rough sets vii, Springer-Verlag, Berlin, Heidelberg, chapter Applications of Rough Set Based K-means, Kohonen SOM, GA Clustering, pp. 120–139.

Liu, K.-Y., Zang, T. and Wang, L. (2010). A new parallel video understanding and retrieval system, *IEEE ICME*, Suntec City, pp. 679–684.

Liu, N., Wu, H. and Lin, L. (2015). Hierarchical ensemble of background models for ptz-based video surveillance, *IEEE Trans. Cyberns.* **45**(1): 89–102.

Liu, W., Anguelov, D., Erhan, D., Szegedy, C., Reed, S., Fu, C.-Y. and Berg, A. C. (2016a). Ssd: Single shot multibox detector, *European conference on computer vision*, Springer, pp. 21–37.

Liu, Y., Xie, H., Wang, L. and Tan, K. (2016b). Hyperspectral band selection based on a variable precision neighborhood rough set, *Appl. Opt.* **55**(3): 462–472.

Maddalena, L. and Petrosino, A. (2008). A self-organizing approach to background subtraction for visual surveillance applications, *IEEE Trans. Image Process* **17**: 1168–1177.

Maggio, E. and Cavallaro, A. (2010). *Video Tracking—Theory And Practice*, Wiley, West Sussex, UK.

Meher, S. K. and Pal, S. K. (2011). Rough-wavelet granular space and classification of multispectral remote sensing image, *Appl. Soft Comput.* **11**: 5662–5673.

Milan, A., Roth, S. and Schindler, K. (2014). Continuous energy minimization for multitarget tracking, *IEEE Trans. PAMI* **36**(1): 58–72.

Milbich, T., Bautista, M., Sutter, E. and Ommer, B. (2017). Understanding videos, constructing plots learning a visually grounded storyline model from annotated videos, *IEEE ICCV*, Venice, Italy, pp. 4404–4414.

Mottaghi, R., Fidler, S., Yuille, A., Urtasun, R. and Parikh, D. (2016). Human-machine crfs for identifying bottlenecks in scene understanding, *IEEE Trans. PAMI* **38**(1): 74–87.

Mushrif, M. M. and Ray, A. K. (2008). Color image segmentation: Roughset theoretic approach, *Pattern Recogn. Lett.* **29**(4): 483–493.

Nawaz, T., Poiesi, F. and Cavallaro, A. (2014). Measures of effective video tracking, *IEEE Trans. Image Proc.* **23**(1): 5–43.

Ngu, A. H., Gutierrez, M., Metsis, V., Nepal, S. and Sheng, Q. Z. (2017). Iot middleware: A survey on issues and enabling technologies, *IEEE Internet Things J.* **4**(1): 1–20.

Nguyen, H. T. and Smeulders, A. W. M. (2004). Fast occluded object tracking by a robust appearance filter, *IEEE Trans. PAMI* **26**: 1099–1104.

Pal, S. K. and Meher, S. K. (2013). Natural computing: A problem solving paradigm with granular information processing, *Appl. Soft Comput.* **13**(9): 3944–3955.

Pal, S. K., Meher, S. K. and Skowron, A. (2015). Data science, big data and granular mining, *Pattern Recognit. Lett.* **67**: 109–112.

Pal, S. K., Shankar, B. U. and Mitra, P. (2005). Granular computing, rough entropy and object extraction, *Pattern Recogn. Lett.* **26**(16): 2509–2517.

Pawlak, Z. (1992). *Rough Sets: Theoretical Aspects of Reasoning about Data*, Kluwer Academic Publishers, Norwell, MA.

Pawlak, Z. (1998). Granularity of knowledge, indiscernibility and rough sets, *In Proc., IEEE International Conference on Fuzzy Systems*, pp. 106–110.

Pedrycz, W. and Song, M. (2014). A granulation of linguistic information in ahp decision-making problems, *Inf. Fusion* **17**: 93–101.

Pernici, F. and Bimbo, A. (2014). Object tracking by oversampling local features, *IEEE Trans. PAMI* **36**(12): 2538–2551.

Peters, J. F. and Borkowski, M. (2004). K-means indiscernibility relation over pixels, S. Tsumoto, R. Slowinski, J. Komorowski, J.W. Grzymala-Buess, *Lecture Notes in Artificial Intelligence*, Springer-Verlag, Berlin, pp. 580–535.

Redmon, J., Divvala, S., Girshick, R. and Farhadi, A. (2016). You only look once: Unified, real-time object detection, *Proceedings of the IEEE conference on computer vision and pattern recognition*, pp. 779–788.

Ren, S., He, K., Girshick, R. and Sun, J. (2015). Faster r-cnn: Towards real-time object detection with region proposal networks, *Adv. Neural Inf. Process Syst.* pp. 91–99.

Schmidhuber, J. (2015). Deep learning in neural networks, *Neural Netw.* **61**(C): 85–117.

Sen, D. and Pal, S. K. (2009a). Generalized rough sets, entropy, and image ambiguity measures, *IEEE Trans. Syst. Man Cybern. Part B* **39**(1): 117–128.

Sen, D. and Pal, S. K. (2009b). Histogram thresholding using fuzzy and rough measures of association error, *IEEE Trans. Image Proc.* **18**(4): 879–888.

Serby, D., Koller-Meier, E. and Gool, L. J. V. (2004). Probabilistic object tracking using multiple features, *17th International Conference on ICPR 2004, Cambridge, UK, August 23–26, 2004*, pp. 184–187.

Shen, C., Kim, J. and Wang, H. (2010). Generalized kernel-based visual tracking, *IEEE Trans. Circ. Syst. Vid. Technol.* **20**: 119–130.

Shum, H. P. H., Ho, E. S. L., Jiang, Y. and Takagi, S. (2013). Real-time posture reconstruction for microsoft kinect, *IEEE Trans. Cyberns.* **43**(5): 1357–1369.

Skowron, A. and Rauszer, C. (1992). The discernibility matrices and functions in information systems, *in* R. Slowiński (ed.), *Intelligent Decision Support, Handbook of Applications and Advances of the Rough Sets Theory*, Kluwer Academic, Dordrecht.

Smeulders, A., Chu, D., Cucchiara, R., Calderara, S., Dehghan, A. and Shah, M. (2014). Visual tracking: An experimental survey, *IEEE Trans. PAMI* **36**(7): 1442–1468.

Stauffer, C. and Grimson, W. E. L. (1999a). Adaptive background mixture models for real-time tracking, *Computer Vision and Pattern Recognition, 1999. IEEE Computer Society Conference on.*, Vol. 2, IEEE, pp. 246–252.

Stauffer, C. and Grimson, W. E. L. (1999b). Adaptive background mixture models for real-time tracking, *IEEE CVPR*, pp. 246–252.

Stauffer, C. and Grimson, W. E. L. (2000). Learning patterns of activity using real-time tracking, *IEEE Trans. PAMI* **22**: 747–757.

Swirniaski, R. W. (2001). Rough sets methods in feature reduction and classification, *Int. J. Appl. Math. Comp. Sci.* **11**: 565–582.

Thiesse, F., Floerkemeier, C., Harrison, M., Michahelles, F., Roduner, C. (2009). Technology, standards, and real-world deployments of the epc network, *IEEE Internet Comput.* **13**(2): 36–43.

Veenman, C. J., Reinders, M. J. T. and Backer, E. (2001). Resolving motion correspondence for densely moving points, *IEEE Trans. PAMI* **23**(1): 54–72.

Wang, Q., Chen, F., Xu, W. and Yang, M.-H. (2012). Object tracking via partial least squares analysis, *IEEE Trans. Image Proc.* **21**(10): 4454–4465.

Wójcik, Z. M. (1994). Application of rough sets for edge enhancing image filters, *Proceedings 1994 International Conference on Image Processing, Austin, Texas, USA, November 13–16, 1994*, pp. 525–529.

Wu, Y., Lim, J. and Yang, M. (2015). Object tracking benchmark, *IEEE Trans. PAMI* **37**(9): 1834–1848.

Xia, F., Yang, L. T., Wang, L. and Vinel, A. (2012). Internet of things, *Int. J. Commun. Syst.* **25**(1): 1101–1102.

Xiao, J. and Oussalah, M. (2016). Collaborative tracking for multiple objects in the presence of inter-occlusions, *IEEE Trans. Circ. Syst. Vid. Technol.* **26**(2): 304–318.

Yang, Y., Liu, J. and Shah, M. (2009). Video scene understanding using multi-scale analysis, *IEEE Intl. Conf. Comp. Vision*, Kyoto, pp. 1669–1676.

Yao, J. T., Vasilakos, A. V. and Pedrycz, W. (2013). Granular computing: Perspective and challenges, *IEEE Trans. Cyberns.* **43**(6): 1977–1989.

Yao, Y. (2011). Two semantic issues in a probabilistic rough set model, *Fundam. Inform.* **108**: 249–265.

Yao, Y. Y. (2000). Granular computing: basic issues and possible solutions, *Proceedings of the 5th Joint Conference on Information Sciences*, pp. 186–189.

Yilmaz, A., Javed, O. and Shah, M. (2006). Object tracking: A survey, *ACM Comput. Surv.* **38**(4): 1264–1291.

Zadeh, L. A. (1997). Toward a theory of fuzzy information granulation and its centrality in human reasoning and fuzzy logic, *Fuzzy Sets Syst.* **90**(2): 111–127.

Zaidenberg, S., Boulay, B. and Bremond, F. (2013). A generic framework for video understanding applied to group behavior recognition, *IEEE AVSS*, Beijing, pp. 136–142.

Zhang, J., Marszaek, M., Lazebnik. S. and Schmid, C. (2007). Local features and kernels for classification of texture and object categories, *IJCV* **73**(2): 259–289.

Zhang, K., Nanjing, C., Zhang, L. and Yang, M.-H. (2014). Fast compressive tracking, *IEEE Trans. PAMI* **36**(10): 2002–2015.

Chapter 2

Partial Supervised Tracking

2.1 Introduction

This chapter of the book demonstrates the effectiveness of rough set theoretic granular computing in the task of partial supervised tracking, that is tracking with initial manual interactions. Detection of the moving object(s) as a separate segment in the frame is the first objective for the task of tracking. The object(s) can be separated out from its background according to spatial homogeneity and/or temporal homogeneity. If the object and background are separable in any feature space, then only spatial segmentation is sufficient to give satisfactory result. But, in most of the cases this is not true, and the total object cannot be solely extracted out from the background using spatial segmentation. On the other hand, temporal information plays an important role in detecting objects. But, without huge change from frame-to-frame or without having a predefined background, temporal segmentation technique is also unable to extract the total object out in the video sequence (Yilmaz *et al.*, 2006). There exist a number of methods where tracking of objects in a video sequence starts with tagging those objects manually in the initial frames. Searching of those particular objects in the following frames executes the task of tracking.

In the following sections, we describe in detail a method demonstrating how the concept rough set theoretic granular computing can be used in partial supervised tracking. Before that we discuss some of the popular benchmark approaches of such tracking in Section 2.2.

2.2 Popular Approaches for Partially Supervised Tracking

Here we mention some popular methods, where tracking by matching among different segments in consecutive video frames is the primary task. Briechle and Hanebeck (2001) proposed a most basic concept of tracking through direct target matching by normalized cross-correlation which uses the intensity values in the initial target bounding box as template. Baker and Matthews (2004) developed a tracker that finds the affine-transformed match between the target bounding box and candidate windows around the previous location. Nguyen and Smeulders (2004) addressed the issue of appearance change by appearance-predicted matching for handling target change under occlusion. The famous mean-shift tracker (Comaniciu *et al.*, 2000) performs matching with histograms rather than using any spatial information about the pixels, making it suitable for radical shape changes of the target. Ross *et al.* (2008) developed a tracker to keep an extended model of appearances for capturing the full range of appearances of the target in the past. Kwon and Park (2009) designed a tracker with extended model of appearances which extends the traditional (translation, scale, rotation) motion types to a more general 2-D affine matrix group. Chehovin *et al.* (2011) aimed to detect rapid and significant appearance changes with sparse optimization in two layers. The changed regions are mapped in the local layers by maintaining a global layer in this method. In another approach, namely, L1 (Manhattan Distance)-minimization tracker proposed by Mei and Ling (2009), sparse optimization by L1 from the previous frames is employed. It starts using the intensity values in target windows sampled near the target as the bases for a sparse representation. Individual non-target intensity values are used as alternative bases. Candidate windows in the new frame are sampled from a Gaussian distribution centered at the previous target position by particle filtering. Hu *et al.* (2015) defined a technique where the templates from different image features are fused to form a robust object model and the tracking is done with particle filter. Park *et al.* (2015) proposed many to one (M to 1) optimization approach for tracking multiple interacting objects. Milan *et al.* (2014) developed an energy-based model of multi-target tracking for all the target locations and all video frames in a given time window. Object tracking is formulated as an optimization problem by considering both the reconstruction and classification errors in the objective function by Wang *et al.* (2015). Kwon and Lee (2014) came up with a method where several samples of the states of the target and the trackers are considered together during the sampling process using Markov Chain Monte Carlo (MCMC) method. Here the trackers interactively communicate and exchange information with others; thereby improving the overall tracking performance. All the aforesaid methods require initial manual interaction. Therefore, these can be said to be partially supervised methods.

Let us now discuss in the following sections about a rough entropy-based partial supervised tracking method, namely, RE-SpTmp with experimental results (Chakraborty *et al.*, 2013). It uses spatio-temporal segmentation technique for tracking, where the spatial and temporal segmentation outputs are merged together to construct the target object for its detection and tracking efficiently and accurately. The spatial segmentation over each frame of a sequence is done with rough set theoretic granular computing. The temporal segmentation is done afterwards based on beta distribution of three consecutive frames. The results of these two segmentations are merged and the moving object(s) are tracked. Rough entropy (discussed in details in Section 1.4.8) (Pal *et al.*, 2005) with unequal image granulation is used incorporating some modifications for video image segmentation. The basic idea underlying the rough set theoretic approach to information granulation is to discover the extent to which a given set of objects (e.g., pixels in windows of an image) approximates another object of interest (e.g., image region). Here the temporal segmentation is carried out with *Beta Distribution* (Donald and Samuel, 1983; Lazo and Rathie, 1978).

Detection of an object can be viewed in another way as removal of background. In case of video sequences, the proper estimation of a background plays an important role. There are several methods using frame difference, median filtering, Kalman filtering, Mixture of Gaussian, self organizing approach, etc. for background construction (Cucchiara *et al.*, 2003; Maddalena and Petrosino, 2008; Stauffer and Grimson, 1999). The time complexity or memory requirements of most of these techniques is very high. We know that when a random variable follows *Beta Distribution* (Donald and Samuel, 1983; Lazo and Rathie, 1978), the mean and standard deviation of the variable can be estimated by three-point estimation (TPE). It is normally

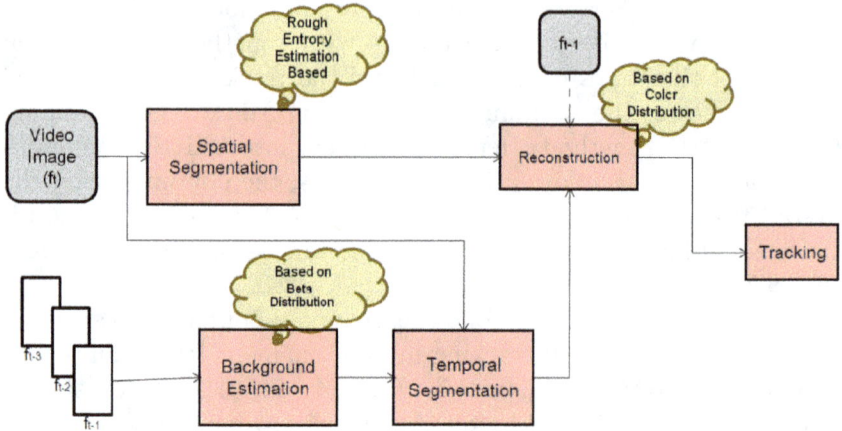

Fig. 2.1: Block diagram of Rough Entropy-based Spatio-Temporal Tracking (RE-SpTmp).

used in time estimation in management issues (Malcolm *et al.*, 1959). This three-point estimation of Beta Distribution is used in the aforesaid RE-SpTmp for temporal segmentation.

The block diagram of the RE-SpTmp is given in Fig. 2.1, where the bubbles show the methodologies that we are going to discuss in detail and that are used for the tasks in the rectangular blocks.

Please note from Fig. 2.1 it is seen that RE-SpTmp method involves two parts, those are spatial segmentation and temporal segmentation. The spatial segmentation is carried out with rough entropy maximization with unequal granules (RE-UnEqGr) and the temporal segmentation is done with three point estimation (TPE).

The rest of this chapter is organized as follows. In Sec. 2.3, we have discussed the basic concepts of a bi-level spatial segmentation technique with rough entropy maximization (Pal *et al.*, 2005; Uma Shankar and Chakraborty, 2011; Chakraborty *et al.*, 2013), namely, RE-UnEqGr. In Section 2.4, technique of background estimation based on a three-point estimation is described. In Section 2.5, both the segmentation results obtained after spatial and temporal segmentation are combined, based on the statistical distribution of their intersection, along with object movement information and an algorithm for object tracking in video sequence. In Section 2.6.1, we have shown some results of spatial segmentation, background estimation, object reconstruction and tracking. First, we have provided comparative results of rough entropy maximization-based spatial segmentation technique on several types of video images with some other well-known spatial segmentation techniques (e.g., Otsu's thresholding, rough entropy maximization with uniform granules (Pal *et al.*, 2005) and the rough fuzzy entropy based segmentation (Sen and Pal, 2009). Their quantitative comparison is also made in terms of β index (Pal *et al.*, 2000) and DB index (Davies and Bouldin, 1979). Next, the comparative performance of temporal segmentation is shown with a popular and widely used technique: mixture of Gaussian (MoG) and a change detection technique: linear change detection (LDD) (Durucan and Ebrahimi, 2001). All these demonstrate how RE-SpTmp results in less noisy foreground. Finally, the results of reconstruction and tracking are provided. The reconstruction results were validated with ground truth. The object(s) in the sequences has (have) been found to be successfully tracked.

2.3 Rough Entropy-based Spatial Segmentation

Here the image is defined as a rough set following the definition of Section 1.4.7. And the rough entropy is measured according to Section 1.4.8. To detect the object, we describe here the method of object enhancement/extraction based on the principle of minimizing the roughness of both object and background regions, i.e., maximizing RE_T. One can compute for every T the RE_T of the image, representing the background and object regions $(0, \ldots, T)$ and $(T + 1, \ldots, L - 1)$, respectively, and select the one for which RE_T is maximum. That level is viewed as the optimum threshold to provide the object–background segmentation. Note that maximizing the rough entropy to get the required threshold basically implies minimizing both the object *roughness* and background *roughness*.

2.3.1 *Formation of Granules*

A granule is a clump of objects (points), in the universe of discourse, drawn together by indistinguishability, similarity, proximity or functionality (Zadeh, 1997). There exist several methods for the formation of a granule for measuring the ambiguity in images (Sen and Pal, 2009; Yao, 2000). In RE-UnEqGr, the formation of unequal granules is made according to that discussed in Uma Shankar and Chakraborty (2011), Pal and Chakraborty (2013), Chakraborty *et al.* (2013). The granules are formed by drawing the pixels of an image together based on their spatial adjacency, as well as gray level similarity. The technique used is called quadtree decomposition of each frame of the video sequence. Here, each frame is decomposed (split) into four quadrants, if the difference between maximum and minimum gray values is greater than a specified threshold $GrTh$, say, each sub-quadrant, thus produced, is again checked for further decomposition by the same criterion. The process is repeated till there is no further split. The quadrants (regions) so generated finally result in what is called unequal granules. For example, one may go through Fig. 1.4, where the granules generated after performing quadtree decomposition on the image are shown with white partitions. This is how both the gray level and spatial similarities are taken care of while forming the granules. The first and the third quartiles of the image gray level distribution (denoted as Q_1 and Q_3) are considered to calculate the said threshold ($GrTh$) for granule detection as

$$GrTh = \frac{Q_3 - Q_1}{2}. \qquad (2.1)$$

Here image statistics is used for the choice of the threshold ($GrTh$) for determining granules automatically. The resulting granules are unequal in size, which is more appropriate and natural for real-life problems.

Note: In RE-UnEqGr, the quad-tree decomposition is done only once on a frame and the same decomposition is considered for the consecutive frames. Moreover, the threshold detection considering all possible values of gray level is also carried out only once, and in the consecutive frames the search for the optimal threshold (T^*) is limited only around that obtained in the previous frame. These two implementation strategies improve the speed of detection of the object without deterioration in accuracy.

Although gray level is used here as the feature for granule formation, one may consider some other feature like color, texture or their combination depending on the application.

2.3.2 Algorithm for Optimal Thresholding for Spatial Segmentation

Following is the algorithm for efficient implementation of the aforesaid methodology for selecting the optimal threshold T^* for spatial segmentation.

Let max_gray and min_gray be the maximum and minimum gray level values of the image, respectively. Let the ith granule $(granule_i)$ represent a window of $m_i \times n_i$ pixels. Let the total number of granules be $total\_no\_granule$ and the size of the ith granule be $size\_of\_granule_i = m_i \times n_i$.

Initialize: Four integer arrays, namely, $object\_lower$, $object\_upper$, $background\_lower$, $background\_upper$ each of size $(max\_gray - min\_gray + 1)$ to zero.

Step 1: For $i = 1$ to $total\_no\_granule$
$max\_granule_i$ = maximum gray value of pixels in $granule_i$
$min\_granule_i$ = minimum gray value of pixels in $granule_i$

(a) for $max\_granule_i \leq j \leq max\_gray$
$object\_lower(j) = object\_lower(j) + size\_of\_granule_i$

(b) for $min\_granule_i \leq j \leq max\_gray$
$object\_upper(j) = object\_upper(j) + size\_of\_granule_i$

(c) for $min\_gray \leq j \leq min\_granule_i$
$background\_lower(j) = background\_lower(j)$
$+ size\_of\_granule_i$

(d) for $min\_gray \leq j \leq max\_granule_i$
$background\_upper(j) = background\_upper(j)$
$+ size\_of\_granule_i$

Step 2: For $l = min\_gray$ to max_gray
$object\_roughness(l) = 1 - [object\_lower(l)/object\_upper(l)]$
$background\_roughness(l) = 1 -$
$[background\_lower(l)/background\_upper(l)]$
IF $object\_roughness(l) \leq \frac{1}{(BASE)}$, then $object\_entropy(l) = 1$
(i.e., noise is removed),
ELSE $object\_entropy(l) = [object\_roughness(l)$
$log_{BASE}(object\_roughness(l))]$, AND
IF $background\_roughness(l) \leq \frac{1}{(BASE)}$, then
$background\_entropy(l) = 1$ (i.e., noise is removed),

ELSE *background_entropy(l)* =
$\qquad$ [*background_roughness(l)*
log_{BASE} (*background_roughness(l)*)]
Rough_entropy(l) = $-[\frac{BASE}{2}]$ × [*object_entropy(l)* + *background_entropy(l)*].

Step 3: *Threshold* T^*(optimal) = arg $\max\limits_{l}$ [*rough_entropy(l)*]. ♠

2.4 Background Estimation-based Temporal Segmentation

The aim of this background estimation technique is to construct a background model from the available information of the video sequence so that the object can easily be separated. This process is also known as temporal segmentation, that is extracting the moving pixels from a video as a segment.

There are several methods to estimate the background of a video sequence (Cheung and Kamath, 2005). Frame difference, median filter, linear predictive filter, approximate median filter, Kalman filter, mixture of Gaussian are the commonly used techniques to construct a background of a video sequence taken from a still camera. In RE-SpTmp, three-point approximation-based background estimation (TPE) is used.

2.4.1 *Three-point Estimation Technique*

In this method of temporal segmentation, the statistical distribution of gray levels for one pixel in previous N frames is taken into account. It can be said that the video sequence is a discrete probability distribution of gray levels for every pixel. According to (Donald and Samuel, 1983), when a random variable follows *Beta Distribution*, the mean and standard deviation of the variable can be estimated by three-point estimation. This technique is used to estimate the background. The standard deviation of distribution is considered for possible error due to randomness. Two parameters (i.e., mean and standard deviation) together enable in characterizing the background. The estimated background is subtracted from the target frame to delineate the object.

The beta distribution along with triangular distribution (Donald and Samuel, 1983) is used to model events (Chakraborty *et al.*, 2013) which are constrained to take place within an interval defined by an optimistic and a pessimistic value. In PERT (Program Evaluation and Review Technique) (Donald and Samuel, 1983) this property of beta distribution was extensively used. The technique evolved around the approximations for the mean and standard deviation of a Beta distribution. Here the three points used to estimate the mean and the standard deviation are named as optimistic point, most likely point and pessimistic point. The estimation of mean and variance is done as follows:

$$\mu(X) = \frac{a + 4b + c}{6} \tag{2.2}$$

$$\sigma(X) = \frac{a - c}{6}. \tag{2.3}$$

In Eq. (2.2) a, b and c denote the optimistic, most likely and pessimistic points, respectively. The above approximations provide a good estimate for mean and standard deviation of the distribution.

Suppose the previous three frames (of size $M \times N$) are used to estimate the background and standard deviation. Let the previous three frames of the tth frame be denoted by f_{t-1}, f_{t-2} and f_{t-3}, respectively. Then the estimation of the background, based on the aforesaid three-point distribution, is characterized by

$$a = max(f_{t-1}, f_{t-2}, f_{t-3}) \tag{2.4}$$

$$b = median(f_{t-1}, f_{t-2}, f_{t-3}) \tag{2.5}$$

$$c = min(f_{t-1}, f_{t-2}, f_{t-3}). \tag{2.6}$$

a, b and c denote the optimistic, most likely and pessimistic values of estimation, respectively. The mean and standard deviation of these estimations are defined accordingly as follows:

$$\widehat{f_t} = \frac{a + 4b + c}{6} \tag{2.7}$$

$$\widehat{\sigma_t} = \frac{a - c}{6}. \tag{2.8}$$

In Eq. (2.7), $\widehat{f_t}$ is the estimated background corresponding to frame f_t. A pixel $f_t(x, y)$ will be detected as a foreground pixel if its difference from the estimated value is greater than E (normally chosen as 3, as it is found experimentally suitable) times of variance. Therefore,

(x, y) will be a foreground pixel if

$$|f_t(x, y) - \widehat{f}_t(x, y)| > E\widehat{\sigma}_t. \tag{2.9}$$

Otherwise, the pixel will be treated as a background pixel. The aforesaid way of selecting a, b and c is justified, because in a video sequence, it is natural to assume that the median values will be the most likely ones to be the background, and the fraction of the difference between maximum and minimum values of a distribution will be helpful to find the variance of pixels. Therefore, such an approximation can give a better estimate of the background. It is also fast to approximate.

Note: Three points (a, b, c) have been derived here from the previous three frames. However, one can also use more frames.

2.5 Target Localization and Tracking: Algorithm

The object detection and tracking starts with the selection of a tracker. Suppose, a rectangular tracker is considered, which completely covers the object that is to be tracked. The initial tracker is defined as a rectangular box, whose vertices are given as $((T_{XL}, T_{YL})$, (T_{XU}, T_{YL}), (T_{XU}, T_{YU}) and $(T_{XL}, T_{YU}))$, shown in Fig. 2.2. Then the following are the steps to detect the object and to track it. In the beginning of the algorithm, the first frame is considered as the reference frame. Then it is segmented into regions and the object of interest is marked.

Step 1: Convert the input color (RGB) image (I) to gray level image (Y) by using the equation: $Y = 0.3R + 0.59G + 0.11B$.

Step 2: Apply the rough entropy-based image segmentation method as described in Section 2.3. (*Note*: Here, the merit of considering a window around the threshold in the previous frame is exploited for segmentation of video images in the current frame).

Step 3: Apply the temporal segmentation method according to three-point estimation (Section 2.4.1).

Step 4: Design the tracker according to the following criteria:
IF $d_x > 0$, then $T_{xu} = T_{xu} + 1.25s$ and $T_{xl} = T_{xl} - 0.75s$
ELSE $T_{xu} = T_{xu} + 0.75s$ and $T_{xl} = T_{xl} - 1.25s$

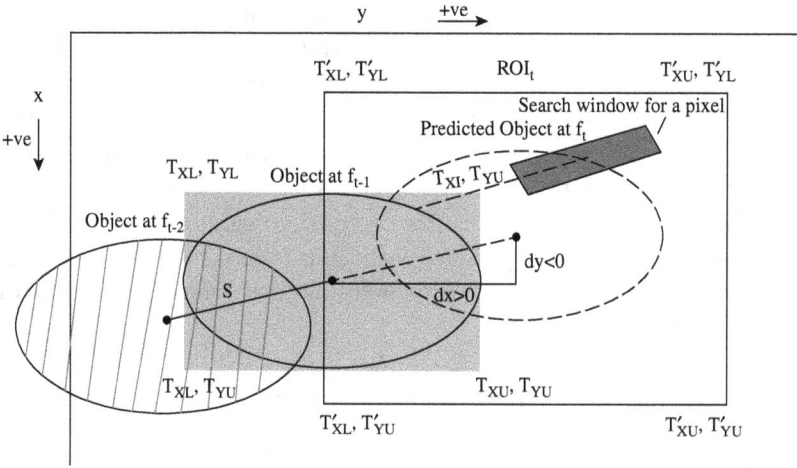

Fig. 2.2: Pictorial representation of the defined tracker with labeled vertices.

IF $d_y > 0$, then $T_{yu} = T_{yu} + 1.25s$ and $T_{yl} = T_{yl} - 0.75s$
ELSE $T_{yu} = T_{yu} + 0.75s$ and $T_{yl} = T_{yl} - 1.25s$.
Here, $T_x = T_{xu} + T_{xl}$ and $T_y = T_{yu} + T_{yl}$.

Step 5: Consider the intersection of the two segmented images within the tracker in RGB feature space and compute the mean (mn) and maximum deviation (max_dev) of those points in the same feature space.

Step 6: For every pixel (x, y) within the tracker, which belongs to either the spatial or temporal segmented region, check the following:
IF $|f_R(x,y) - mn| < max\_dev_R$ and $|f_G(x,y) - mn| < max\_dev_G$ and $|f_B(x,y) - mn| < max\_dev_B$, then $F(x,y) = I(x,y)$ ELSE $F(x,y) = 0$ (here, F is the detected object image).

Step 7: Redefine the tracker around the detected object for tracking.

Step 8: Repeat Steps 1 to 8, for all the frames in the video sequence for tracking of the moving object. ♠

2.6 Experimental Results

Here we present some experimental results (Chakraborty *et al.*, 2013) along with comparisons to evaluate the effectiveness of RE-SpTmp in

(a) spatial segmentation, (b) temporal segmentation, (c) reconstruction and (d) tracking. Different types of datasets used are: (1) *Surveillance Scenario* from PETS-2000 [PETS-2000 (2000)], (2) *Walk3* from PETS-2004 [PETS-2004 (2004)] and (3) *5b* from OTCBVS-2007 (Davis and Sharma, 2007). Performance was studied almost over 2000 frames in total, out of which a few are shown here.

At first we present the results of spatial segmentation and temporal segmentation, and demonstrate their comparative performance with other existing methods. This is followed by the results of reconstruction and tracking.

2.6.1 *Results of Spatial Segmentation*

In Fig. 2.3, we have shown a comparative study of RE-UnEqGr with different bi-level segmentation methods. The segmentation is carried out on three frames of the aforesaid video sequences. Here one individual frame of a video sequence is considered as a still image. Comparing methods are Otsu's thresholding, rough entropy maximization-based thresholding (RE) with uniform granules (Pal *et al.*, 2005) of sizes 4×4 and 6×6, and rough-fuzzy entropy-based segmentation (RFE) [Sen and Pal (2009)] with crisp set and crisp granules of 6×6 size. RE-UnEqGr involves unequal granules, as described in Section 2.3.1.

From Fig. 2.3, it can be seen visually that the RE-UnEqGr method can separate out the object of interest more clearly than the other methods. Though rough entropy method (RE) with granules of size 6×6 gives equally good results in some of the cases, the size of the granule is to be chosen experimentally (Pal *et al.*, 2005), which is not always practically feasible.

Figure 2.4 (Chakraborty *et al.*, 2013) demonstrates the aforesaid comparative performance quantitatively in terms of β index (Pal *et al.*, 2000) and DB index (Davies and Bouldin, 1979) for all frames (Frame nos. 133–180) of *Surveillance Scenario* (PETS-2000, 2000) video sequence, as an example. The β index and DB index of a segmented image with C number of classes are defined as follows:

$$\beta = \frac{\sum_{i=1}^{C} \sum_{j=1}^{n_i} (I_{ij} - \overline{I})^2}{\sum_{i=1}^{C} \sum_{j=1}^{n_i} (I_{ij} - \overline{I_i})^2} \qquad (2.10)$$

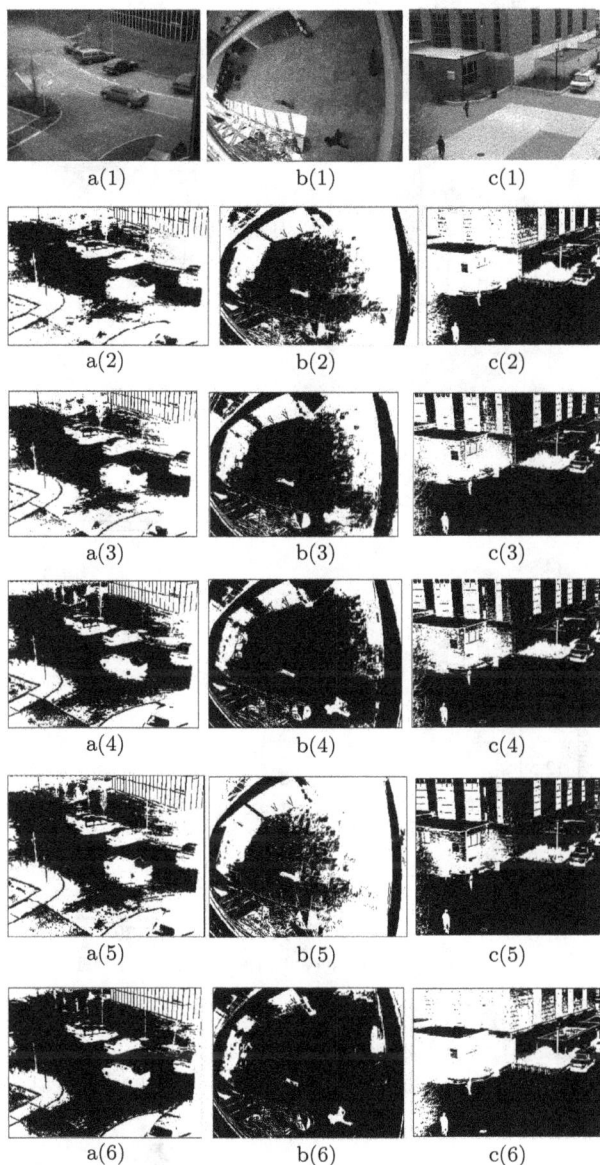

Fig. 2.3: Spatial segmentation results on (a) Frame no. 135 of the *Surveillance Scenario Sequence* from PETS-2000, (b) Frame no. 56 of the *Walk3 sequence* from PETS-2004 and (c) Frame no. 370 of *5b* from OTCBVS-2007 (1) original, (2) Otsu's thresholding, (3) RE with 4×4 granule, (4) RE with 6×6 granule, (5) RFE and (6) RE-UnEqGr.

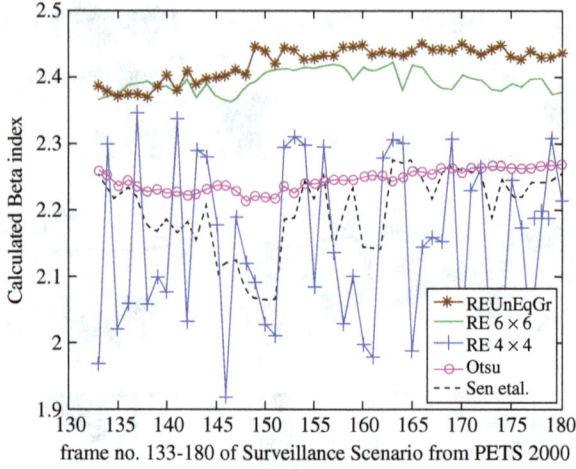

(a)

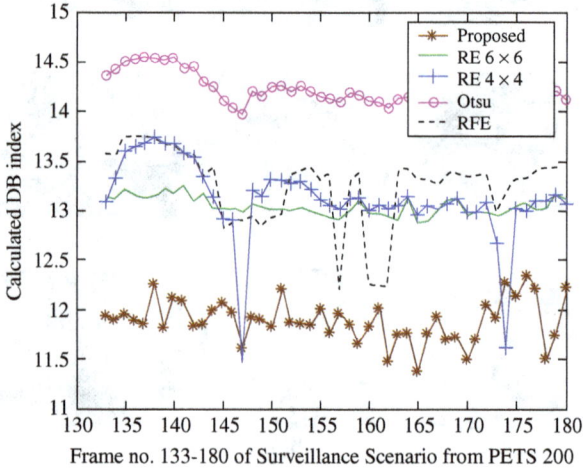

(b)

Fig. 2.4:   Plot of (a) Beta Index (β) and (b) DB Index with frames for Otsu's thresholding, RE with 4×4 and 6×6 granules, RFE and RE-UnEqGr.

$$DB = \frac{1}{C} \sum_{i=1}^{C} \left(max_{i \neq k} \frac{S(v_i) + S(v_k)}{d(v_i, v_k)} \right). \qquad (2.11)$$

Here, in Eq. (2.10), n_i denotes the number of pixels at i^{th} segment in image I. $\overline{I}$ is the mean of the gray values in image I and $\overline{I}_i$ is mean of the gray values in the segment i. In Eq. (2.11), v_i and v_k represent two clusters, $S(.)$ represents the variance and $d(.)$ represents the distance. As seen from Eq. (2.10), for a given number of clusters (segments), the higher value of β and lower value of DB are desirable for good segmentation.

The variation of β values and DB values is shown in Figs. 2.4 (a) and 2.4 (b), respectively. Here the number of classes C is two.

In Fig. 2.4, '*−' line denotes β and DB index obtained by RE-UnEqGr, and '−' and '+−' lines denote the β values and DB values obtained by RE method with 6×6 and 4×4 granules, respectively. The '−' line denotes the values of the two indexes according to RFE, whereas 'o−' denotes the β and DB values obtained after segmentation by Otsu's thresholding. As β-values are higher and DB values are lower in case of RE-UnEqGr, we can say that RE-UnEqGr is also superior quantitatively in the current scenario.

2.6.2   *Results of Temporal Segmentation*

Figure 2.5 (Chakraborty *et al.*, 2013) shows comparative performance of the three-point estimation technique (TPE), with one of the most popular techniques: (MoG) (Stauffer and Grimson, 1999), and a change detection technique: linear change detection (LDD) (Durucan and Ebrahimi, 2001). Results of LDD correspond to 5×5 window, threshold value 0.05 with the frame before the previous frame as the reference frame. In case of TPE, the noise present in the segmented images is much less compared to other two methods, and the object in the current frame can be detected properly as seen from Fig. 2.5. (One may notice that, if an ideal reference frame were available in case of LDD, the results could have been better, but it was not available in the current scenario.) The boundaries of the objects are seen to be clearly extracted with TPE, almost covering the region of interest. This shows that TPE is more efficient and accurate even with less number of frames.

2.6.3   *Results of Reconstruction*

Implementing RE-SpTmp on several video sequences produced satisfactory results for reconstruction and tracking, i.e., the object(s) of interest can be reconstructed properly (very close to the ground truth). In Fig. 2.6, we have shown, as an example, only three of such

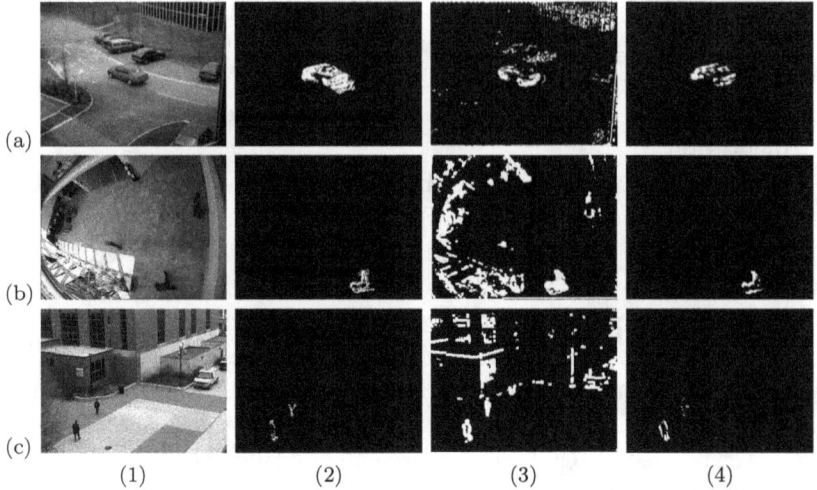

Fig. 2.5:   Results of temporal segmentation by (2) MoG, (3) LDD and (4) TPE method on (a) Frame no. 145 of the *Surveillance Scenario Sequence* from PETS-2000, (b) Frame no. 52 of *Walk3 sequence* from PETS-2004 and (c) Frame no. 448 of *5b* from OTCBVS-2007.

video sequences, one frame from each sequence along with its ground truth. Values of *Precision*, *Recall* and F_{score} for the sequences, thus obtained for evaluation of the performance of reconstruction, are shown in Table 2.1. All the values of *Precision*, *Recall* and F_{score}, as obtained for the different kinds of sequences with single or multiple moving objects are greater than 0.85—signifying high accuracy of the algorithm for reconstruction of the object(s) of interest. Note that, here the ground truth was manually annotated for each frame over every sequence (Chakraborty *et al.*, 2013).

2.6.4   *Results of Tracking*

Here, the results of tracking by implementing the RE-SpTmp algorithm on four frames from each of the aforesaid sequences are shown. Each of the sequences has different types of objects with different kinds of movements. In the first sequence (Fig. 2.7(a)), i.e., the *Surveillance Scenario* from PETS-2000 (PETS-2000, 2000), a car is moving through out the frame and its shape and size are changing gradually. In the second sequence (Fig. 2.7(b)) *Walk3* from PETS-2004 (PETS-2004, 2004), a man is walking through the h-line,

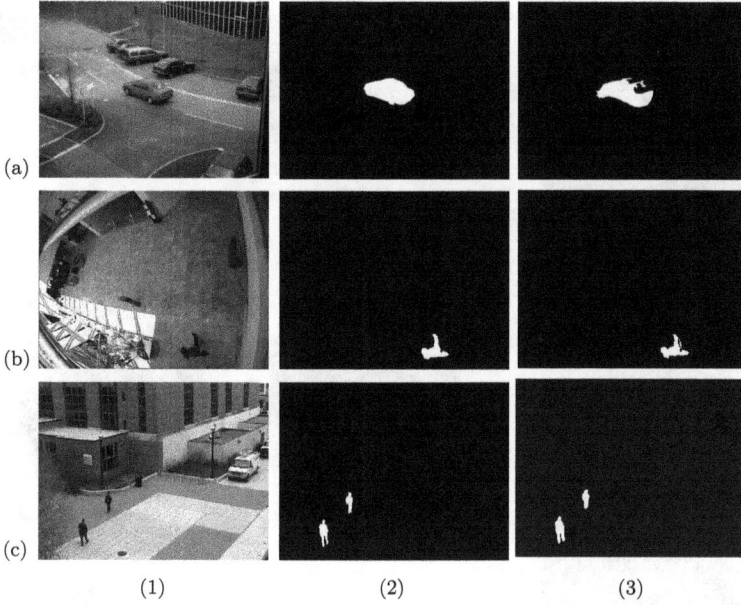

Fig. 2.6: Results of validation (1) original frame, (2) ground truth and (3) reconstruction by RE-SpTmp method of (a) Frame no. 145 of the *Surveillance Scenario Sequence* from PETS-2000, (b) Frame no. 52 of the *Walk3 sequence* from PETS-2004 and (c) Frame no. 448 of *5b* from OTCBVS-2007.

Table 2.1: Reconstruction accuracy on *Surveillance Scenario*, *Walk3* and *5b* sequences.

Sequences	*Recall*	*Precision*	F_{score}
Surveillance Scenario	0.85	0.86	0.85
Walk3	0.89	0.94	0.91
5b	0.86	0.95	0.90

getting stopped, waving his hand and then returning back. There are two objects moving in different directions in the third sequence (Fig. 2.7(c)) *5b* from OTCBVS-2007 (Davis and Sharma, 2007). In all the sequences, the object(s) are being tracked properly, which shows the effectiveness of the algorithm on videos with different characteristics (Chakraborty *et al.*, 2013).

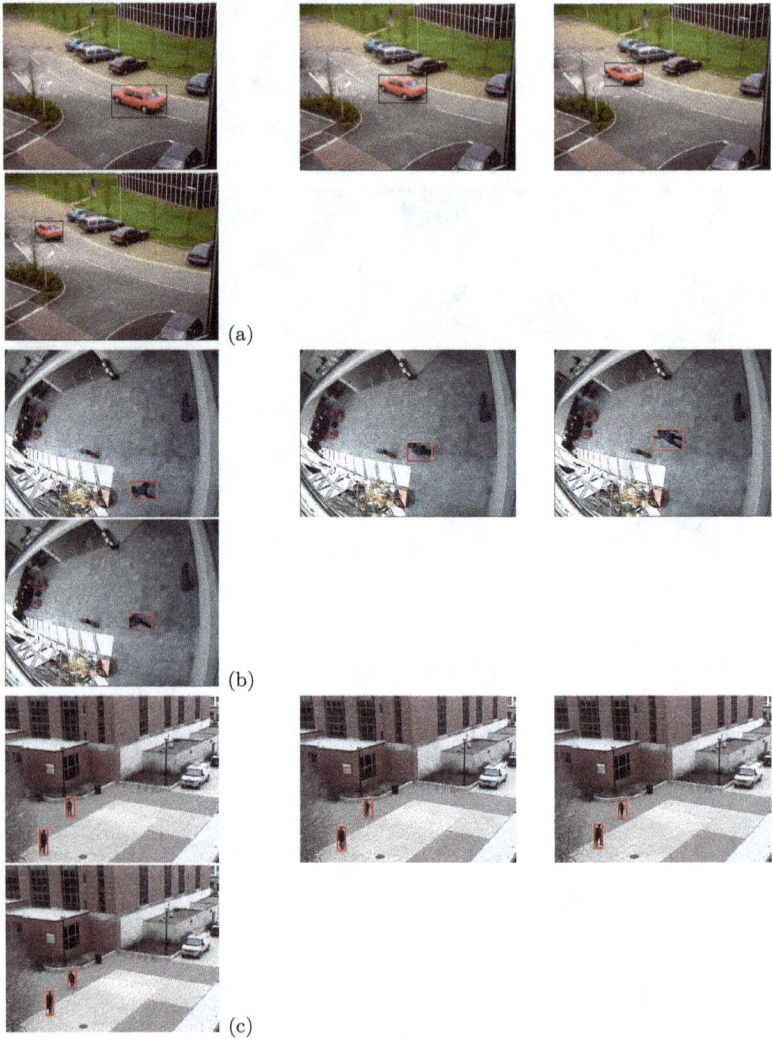

Fig. 2.7: Results of tracking (a) Frame numbers 133, 145, 161, 180 of the *Surveillance Scenario Sequence* from PETS-2000, (b) Frame nos. 58, 87, 101, 159 of *Walk3 sequence* from PETS-2004 and (c) Frame nos. 416, 436, 462, 478 of *5b-sequence* from OTCBVS-2007.

Note that the tracking capability of RE-SpTmp is more robust to gradual illumination change and noise, because the occurrence of such cases in a frame may not affect simultaneously both the spatial and temporal segments of that frame within the search window.

2.7 Conclusions and Discussions

The significance of rough set theoretic granular computing for one kind of object detection, e.g., partially supervised, and tracking from video images is described. Image granulation is performed using quad-tree decomposition resulting in unequal granules, which is closer to natural granulation as compared to equal-sized granules. Here, gray level is used as a feature for formation of granules. One may use any other feature (color, texture, etc. or combination of them) depending on the application.

It is shown that the RE-UnEqGr, based on the aforesaid principle, performs well for segmentation on different types of video images. Results with unequal granules are superior both visually and quantitatively to those of three other techniques involving granules of equal size, as well known Otsu's thresholding. β index and DB index also reflect these well. The temporal segmentation results with TPE are superior to those of MoG and LDD as less amount of noise is present there on the output. The reconstruction results got validated with the ground truths and they have an accuracy of more than 85% for all the sequences under consideration, in some cases it is even as high as 91%. Tracking by RE-SpTmp is also robust to noise and gradual illumination change, as it judiciously combines both spatial and temporal segments for detection.

Overall, this chapter demonstrates the merits of rough set theoretic granular computing for handling the vagueness present in the task of object tracking from video images. The task is considered under partially supervised mode. Here the moving object and background, those from the rough sets, need different gray level distributions and the moving object needs to have similar level variation within it in order to get the same tracked by this method as the algorithm is dependent on spatial segmentation. As a result, the algorithm may fail for tracking multiple objects since the possibility of their belonging to one side of the optimal threshold (T^*) decreases. The next chapter deals with unsupervised tracking of multiple objects where the aforesaid issues concerning threshold selection do not arise. Moreover, the present chapter deals with the concept of lower and upper approximations of rough sets for uncertainty modeling, whereas the next chapter uses the rough set theoretic rule-base with an intelligent adaptation for tracking.

Bibliography

Baker, S. and Matthews, I. (2004). Lucas-kanade 20 years on: A unifying framework, *IJCV* **56**: 221–55.

Briechle, K. and Hanebeck, U. D. (2001). Template matching using fast normalized cross correlation, *in Proc. SPIE* **4387**: 95–102.

Chakraborty, D., Shankar, B. U. and Pal, S. K. (2013). Granulation, rough entropy and spatiotemporal moving object detection, *Appl. Soft Comput.* **13**(9): 4001–4009.

Chehovin, L. Kristan, M. and Leonardis, A. (2011). An adaptive coupled-layer visual model for robust visual tracking, *IEEE ICCV*, Barcelona, Spain.

Cheung, S.-C. S. and Kamath, C. (2005). Robust background subtraction with foreground validation for urban traffic video, *EURASIP J. Appl. Signal Process.* **2005**: 2330–2340.

Comaniciu, D., Ramesh, V. and Meer, P. (2000). Real-time tracking of non-rigid objects using mean shift, *IEEE CVPR*, Hilton Head Island, SC, USA.

Cucchiara, R., Grana, C., Piccardi, M. and Prati, A. (2003). Detecting moving objects, ghosts and shadows in video streams, *IEEE Trans. PAMI* **25**: 1337–1342.

Davies, D. L. and Bouldin, D. W. (1979). A cluster separation measure, *IEEE Trans. PAMI* **1**: 224–227.

Davis, J. and Sharma, V. (2007). Background-subtraction using contour-based fusion of thermal and visible imagery, *Comput. Vis. Image Underst.* **106**: 162–182.

Donald, L. K. and Samuel, E. B. (1983). Three-point approximations for continuous random variables, *Manage. Sci.* **29**: 595–609.

Durucan, E. and Ebrahimi, T. (2001). Change detection and background extraction by linear algebra, *Proc. IEEE* **89**(10): 1368–1381.

Hu, W., Li, W., Zhang, X. and Maybank, S. J. (2015). Single and multiple object tracking using a multi-feature joint sparse representation, *IEEE Trans. PAMI* **37**(4): 816–833.

Kwon, J. and Lee, K. M. (2014). Tracking by sampling and integrating multiple trackers, *IEEE Tran. PAMI* **36**(7): 1428–1441.

Kwon, J. and Park, F. C. (2009). Visual tracking via geometric particle filtering on the affine group with optimal importance functions, *IEEE CVPR*, Miami, FL, USA.

Lazo, A. V. and Rathie, P. N. (1978). On the entropy of continuous probability distributions, *IEEE Trans. Inform. Theory* **IT-24**: 120–122.

Maddalena, L. and Petrosino, A. (2008). A self-organizing approach to background subtraction for visual surveillance applications, *IEEE Trans. Image Proc.* **17**: 1168–1177.

Malcolm, D. G., Roseboom, J. H., Clark, C. E. and Fazar, W. (1959). Application of a technique for research and development program evaluation, *Oper. Res.* **7**: 646–669.

Mei, X. and Ling, H. (2009). Robust visual tracking using l1 minimization, *IEEE ICCV*, Kyoto, Japan.

Milan, A., Roth, S. and Schindler, K. (2014). Continuous energy minimization for multitarget tracking, *IEEE Trans. PAMI* **36**(1): 58–72.

Nguyen, H. T. and Smeulders, A. W. M. (2004). Fast occluded object tracking by a robust appearance filter, *IEEE Trans. PAMI* **26**: 1099–1104.

Pal, S. K. and Chakraborty, D. (2013). Unsupervised tracking, roughness and quantitative indices, *Fundamenta Informaticae (IOS Press)* **124**(1–2): 63–90.

Pal, S. K., Ghosh, A. and Shankar, B. U. (2000). Segmentation of remotely sensed images with fuzzy thresholding, and quantitative evaluation, *Int. J. Remote Sens.* **21**: 2269–2300.

Pal, S. K., Shankar, B. U. and Mitra, P. (2005). Granular computing, rough entropy and object extraction, *Pattern Recogn. Lett.* **26**(16): 2509–2517.

Park, C., Woehl, T. J., Evans, J. E. and Browning, N. D. (2015). Minimum cost multi-way data association for optimizing multitarget tracking of interacting objects, *IEEE Trans. PAMI* **37**(3): 611–624.

PETS-2000 (2000). *IEEE Int. WS Perfor. Evaluation of Tracking and Surveillance.*

PETS-2004 (2004). *IEEE Int. WS Perfor. Evaluation of Tracking and Surveillance and EC Funded CAVIAR project/IST 2001.*

Ross, D. A., Lim, J. and Lin, R. S. (2008). Incremental learning for robust visual tracking, *IJCV* **77**: 125–141.

Sen, D. and Pal, S. K. (2009). Generalized rough sets, entropy, and image ambiguity measures, *IEEE Trans. Syst. Man Cyberns. Part B* **39**(1): 117–128.

Stauffer, C. and Grimson, W. E. L. (1999). Adaptive background mixture models for real-time tracking, *IEEE CVPR*, pp. 246–252.

Uma Shankar, B. and Chakraborty, D. (2011). Spatiotemporal approach for tracking using rough entropy and frame subtraction, *Proceedings of the 4th international conference on Pattern recognition and machine intelligence*, PReMI'11, Springer, pp. 193–199.

Wang, Q., Chen, F., Xu, W. and Yang, M.-H. (2015). Object tracking with joint optimization of representation and classification, *IEEE Trans. Circ. Syst. Video Technol.* **25**(4): 638–650.

Yao, Y. Y. (2000). Granular computing: basic issues and possible solutions, *Proceedings of the 5th Joint Conference on Information Sciences*, pp. 186–189.

Yilmaz, A., Javed, O. and Shah, M. (2006). Object tracking: A survey, *ACM Comput. Surv.* **38**(4): 1264–1291.

Zadeh, L. A. (1997). Toward a theory of fuzzy information granulation and its centrality in human reasoning and fuzzy logic, *Fuzzy Sets Syst.* **90**(2): 111–127.

Chapter 3

Unsupervised Tracking

3.1 Introduction

This chapter demonstrates the merits of another feature of rough sets, namely, rough rule-base for unsupervised tracking of multiple moving objects in a video sequence. A way of forming natural granules over images (frames) without previously defined shape or size is described (Chakraborty and Pal, 2015; Pal and Chakraborty, 2017). The development of a neighborhood granular rough rule-base and adaptation of the rule-base with granular flow graph are also described here for the task of tracking.

According to the existing literature, the rough rule-base is used, so far, over data points only. In the present chapter, the concept of granular rough rule-base is described where both the processing and decision-making are done in granular level. Both the granulation and rough rule-base, though they individually make a process faster, are applied together here for the task of unsupervised tracking over natural neighborhood granulation.

Flow graph was introduced by (Pawlak, 2005) which is a directed acyclic graph used to map information flow. This concept is used in video processing to accumulate information from multiple cameras (Lisowski and Czyzewski, 2014) for behavior modeling. Video tracking involves a decision-making process over time. Incorporation of rough rule-base to this application is proven to be effective with proper updation procedure. The updation of all of the attributes in every frame is time-consuming and reduces the merit of using rule-base. The concept of flow graph is proven to be useful for an intelligent adaptive rule-base generation (Pal and Chakraborty, 2017).

The main advantage of this graph over rule-base is that it can map the information flow between the attributes and hence can show the significance of each individual attribute in an information system. The adaptation can be performed according to the changing significance of the attributes. One way of using flow graph for training and testing is shown in this chapter with two added features (expected flow graph and mutual dependency between flow graphs) (Pal and Chakraborty, 2017). The entire decision-making is performed on granular level, therefore this is called a granular rough flow graph.

The unsupervised rough-rule and flow graph-based approach for tracking, as described here in this chapter, can be broadly divided into three parts: training, testing and updation. Initial P number of frames are given as the input for training, i.e., for the object/background labeling and formation of rough rule-base. The current and its previous frames are the input in testing phase. The 3-D temporal and spatio-color granules are formed over these frames and the object–background separation in the current frame is performed with the rule-base. The rule-base gets updated with the help of flow graph afterwards. The study described in this chapter has the following features. It is shown how (i) spatio-temporal information can be used to form 3-D neighborhood granules in videos with natural partitioning, (ii) neighborhood granular rough rule-base can be formed on 3-D granulated space to reduce the computational complexity, (iii) granular rough flow graph can be defined for rule-base adaptation, and (iv) flow graph can be used for the tasks of testing and validation of the rule-base with two added features, namely, expected flow graph and mutual dependency. All these features characterized the task of unsupervised tracking of moving object(s) with static background, and were proved to be effective in the scenario with no initial manual labeling.

The rest of the chapter is organized as follows. A few popular methods of unsupervised tracking are described in Section 3.2. The formation of 3-D neighborhood granules in NRBFG is described in Section 3.3. The details of formation of rule-base in pixel level and granular level in NRBFG are given in Section 3.4. The method of rule-base adaptaion therein using flow graphs is explained in Section 3.5. This includes definitions of the added features, namely, expected flow graph and mutual dependency between flow graphs. A brief introduction to flow graph and its relevance in tracking is also provided.

Section 3.6 describes the effectiveness of all those features experimentally along with suitable comparisons. Section 3.7 concludes the chapter.

3.2 Popular Methods for Unsupervised Tracking

Unsupervised tracking by computing the difference between consecutive frames has been well studied since the late 1970s (Jain and Nagel, 1979). However, unsupervised tracking with background subtraction became popular with the work of Wren *et al.* (1997). They proposed the method of modeling the color of each pixel of a stationary background with a single Gaussian in LUV (lightness uniform value) feature space to learn the gradual changes in time. The mean and covariance, i.e., the model parameters, are then learned from the color observations in several consecutive frames. Once the background is modeled, the foreground is labeled by measuring the amount of deviation of each pixel from the corresponding background pixel. However, a single Gaussian is not a good model for outdoor scenes as multiple colors can be present in a certain location due to repetitive object motion, shadows or noise. A substantial improvement in background modeling is achieved by using multimodal statistical distributions for background modeling. For instance, Stauffer and Grimson (2000) used a mixture of Gaussians to model the background pixels and the corresponding deviations. In this method, a pixel in the current frame is checked against the background model by comparing it with every Gaussian in the model until a matching Gaussian is found. If a match is found, the mean and variance of the matched Gaussian are updated, otherwise a new Gaussian with the mean equal to the current pixel color and some initial variance is introduced into the mixture. Each pixel is classified based on whether the matched distribution represents the background process. Another approach is to incorporate region-based (spatial) scene information instead of only using color-based information. Elgammal *et al.* (2002) used nonparametric kernel density estimation to model the per-pixel background where the current pixel is matched not only with the corresponding pixel in the background model, but also with those on the nearby pixel locations during the subtraction process; thereby enabling consideration of camera jitter or small movements in the background. Rittsher *et al.* (2000) used Hidden Markov

Models (HMM) to represent the intensity variations of a pixel in an image sequence as discrete states corresponding to the events in the environment to classify small blocks of an image to one of the three (fore-ground, background and shadow) states. Cucchiara *et al.* (2003) defined a method for shadow and ghost (random movement) detection while estimating the background. They used HSI (hue-saturation-intensity) color information for shadow suppression and background model updation. Maddalena and Petrosino (2008) came up with a self-organizing approach through artificial neural networks for background subtraction. This method can handle slightly moving backgrounds or illumination variations. Chi *et al.* (2017) designed a dual deep network for tracking. Its objective is to exploit the hierarchical features in different layers of a deep model and design a dual structure to obtain features from various streams. This model is updated online based on the observation of the tracked object in consecutive frames. Sun *et al.* (2017) developed a Deep Affinity Network (DAN) for tracking multiple objects where object appearance and their affinity between frames were modeled jointly using deep learning. A method of detection, tracking and segmentation of multiple objects with a single convolutional neural network (CNN) was developed by Voigtlaender *et al.* (2019) with dense pixel annotation using two benchmark video datasets. A method of online object tracking and semi-supervised segmentation of the objects was developed by Wang *et al.* (2019) where an improved version of fully-convolutional Siamese network was developed. All of the aforesaid methods work well for tracking multiple moving objects even if the number of total objects is unknown. However, as these methods focused on background modeling, they may fail in detecting overlapping or occlusion in the object regions.

3.3   Neighborhood Granular Rough Rule-base and Flow Graph for Unsupervised Tracking (NRBFG)

A brief overview of the NRBFG method for video tracking is shown in Fig. 3.1. The initial P frames are given to the rule-base for initial labeling, marked as I_1 in the figure. The current frame (f_t) and its previous P frames (f_{t-P}) in a video, marked as I_2, are the input for their processing. The spatio-color and spatio-temporal granules are

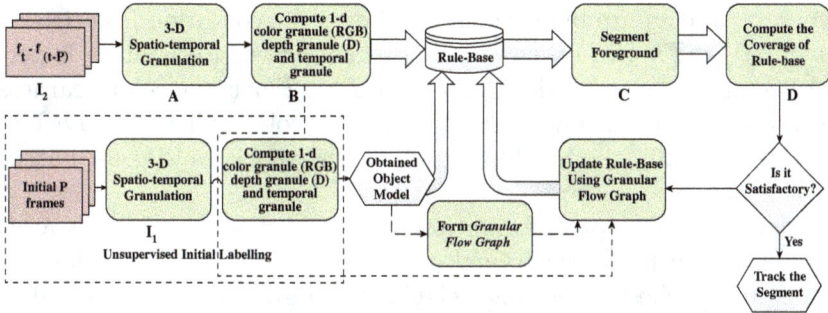

Fig. 3.1: Block diagram of NRBFG tracking method.

formed over these frames, shown in blocks 'A' and 'B', respectively. The decision-making regarding the object–background separation in f_t is performed according to the rule-base. The foreground segmentation is then done (block 'C') depending on the output of the rule-base. The coverage of the rule-base is checked (block 'D') afterwards over the segmented output. If the coverage is satisfactory, then the foreground segment will be tracked, otherwise the rule-base will get updated with the granular flow graph (block 'E').

The details of the tasks of granule formation in blocks 'A', 'B' and I_1 are described in this section. Rule-base and blocks 'C' and 'D' are elaborated in Section 3.4. The block 'E' is described in Section 3.5.

3.3.1 *Formation of Neighborhood Granules*

A meaningful granulation in images and videos was taken into account here. A way of forming spatio-color natural granulation over a still image is explained in the following section. They incorporate spatio-temporal nearness and color nearness of an image under consideration.

3.3.1.1 *Spatio-color Neighborhood Granules*

The spatial nearness and color nearness are incorporated to form spatio-color granules in still images. A granule $\aleph(x_i)$ around a point x_i in universe U is represented as

$$\aleph_{sp-clr}(x_i) = \cup x_j \in U \tag{3.1}$$

where x_i and x_j are binary connected over the condition $|color(x_j) - color(x_i)| < Thr$. In other words, a region growing operation is performed around x_i with threshold value Thr. Thr is the color nearness threshold. That is, how much of a similar color will fall into a bin. In this way only the neighborhood points (spatial and color based) form granules.

Here, the concept of formation of neighborhood granules is a bit different from the conventional one, as in Eq. (1.7). Unlike the previous approaches of forming neighborhood granules as described in Section 1.4.5, neither are the shapes and sizes predefined, nor are the granules formed over every data-point. The granules here are of arbitrary shape and size formed naturally according to the similarities. Further, x_i, around which a granule is formed, is such a point which is not already contained by any other granule. In this way, the formation of unnecessary overlapping granules and complexity can be avoided. Here the overlapping may occur only when the points in the granules have some common neighborhood properties. That is, the overlapping granules are formed depending on the nature of the dataset. Its effectiveness in images is shown in Fig. 3.2 for two images.

It is seen from the figure that the granules (b, c and d of Fig. 3.2) are not of fixed shape or size, rather they represent various

| 1(a) | 1(b) | 1(c) | 1(d) |

| 2(a) | 2(b) | 2(c) | 2(d) |

Fig. 3.2: Granulation over *Lenna* (Top) and *Peppers* (Bottom). (a) Original, (b) $Thr = 50$, (c) $Thr = 30$ and (d) $Thr = 10$.

meaningful segments in images. The main advantages of this granulation over the conventional multilevel segmentation or clustering are that,

- no predefined number of classes is needed;
- no image threshold needs to be defined or found out.

The spatio-color granules, thus formed, contain useful information without keeping all the pixel level information. Dealing with these granules seems to be much more convenient in video processing. These granules are used for decision-making in NRBFG method of tracking as discussed in the following section.

There are two more types of granules used to deal with video sequences in NRBFG. These are 3-D spatio-temporal neighborhood granules and color neighborhood granules. 3-D spatio-temporal neighborhood granules incorporate the temporal nearness which is expected to contain the information on the trajectory of moving object(s) as the third dimension of the granules. The color nearness of these 3-D granules is then considered and color neighborhood granules are formed accordingly. Their formation is described as follows.

3.3.1.2 *3-D Spatio-temporal Neighborhood Granules*

As we mentioned earlier, temporal domain, which is the third dimension of a video stream, makes it different from still images. The processing of temporal information is one of the most important tasks for decision-making in videos. Here a method of extracting this information is described where the highest importance is given to the current frame among the previous P number of frames under consideration instead of giving equal importance to all of them like the other approaches. All the changed information is computed with respect to the current frame. If the current frame (f_t) is of size $M \times N$, and its previous P frames $(f_{t-p} : p = 1, \ldots, P)$ are considered, then the changed information between f_t and F_{t-p} (denoted as τ_p) is computed according to Eq. (3.2). The matrix (τ_p) is of size $M \times N$ and there are P such matrices.

$$\tau_p = |f_t - f_{t-p}| \quad \forall p \in P. \tag{3.2}$$

The median of $\tau_p : p = 1, \ldots, P$ (τ_{med}) is a matrix of size $M \times N$ and is computed as

$$\tau_{med} = Median(\tau_1, \ldots, \tau_P). \tag{3.3}$$

The spatio-temporal granules are formed considering values of the points as τ over the spatial as well as the temporal domains, resulting in three-dimensional granules. Let x_i be the position of a pixel in the current (t^{th}) frame, then the 3-D granules around it are formed according to Eq. 3.4.

$$\aleph_{sp-tmp}(x_i) = \bigcup x_j \in U \tag{3.4}$$

where x_i and x_j are binary connected over $|\tau(x_j) - \tau(x_i)| < Thr_t$ and $x_j \in f_p : p = t, \ldots, t - P$.

The physical interpretation of these 3-D granules is that they contain all the similar changed regions with respect to the current frame in a single granule. In this way, a single granule is expected to contain a single moving object region along with its trajectory.

Note that the value of P depends on the speed of the moving object(s) from frame-to-frame.

3.3.1.3   *Color Neighborhood Granules*

These granules are formed by incorporating the color nearness in a color histogram of a video frame. In a video sequence the moving object models do not usually have the same color values throughout the sequence, rather they deviate. This phenomenon is considered during the granule formation. Besides, the computation with these granules instead of individual color levels is expected to be much faster. These granules are formed with the RGB-D values of the points present in each $\aleph_{sp-tmp}$ (Eq. (3.4)). Let c_i be the color value of a point in $\aleph_{sp-tmp}$. Then the granule around it is defined as

$$\aleph_{clr}(c_i) = \{c_j \in \aleph_{sp-tmp} : \triangle(c_i, c_j) \le Thr_c\}. \tag{3.5}$$

The color and spatio-temporal granules, thus formed from initial P video frames, are used for rule-base generation and adaptation. This adaptive rule-base, formed over the granulated space, is used for tracking unknown input frames.

3.4 Rough Rule-base and Unsupervised Tracking

3.4.1 *Acquiring Features for Initial Unsupervised Labeling*

In case of online processing, the object–background cannot always be labeled manually. Here, one way of object–background estimation is defined. This is different from the conventional unsupervised approaches based on background modeling. Since the computation for background estimation and accordingly the updation processes need to be performed all over the frames, this could be more time-consuming. In NRBFG method for object estimation was adopted, rather the background estimation. Here, all the continuous moving elements in a sequence are treated as the object(s) and the rest as part of the background. The common moving regions of τ_p over P frames may be considered to estimate the lower approximation of the objects ($\underline{O}$), that is,

$$\underline{O} = \bigcap \{\tau_p \quad \forall p = 1, \ldots, P\}. \tag{3.6}$$

The upper approximation of the object(s) ($\overline{O}$) may similarly be modeled over the union of the changed regions of $\underline{O}$ and τ_P (difference between t^{th} and $(t - P)^{th}$ frames). That is,

$$\overline{O} = \bigcup_w \{\underline{O}, \tau_P\}. \tag{3.7}$$

The values of the features that are contained in the set $\underline{O}$ are treated as the core values of the object model, whereas those in the set $\{\overline{O} - \underline{O}\}$, i.e., the boundary region, determine the extent to which the values in the object model are allowed. The values of the attributes in the rule-base that are acquired from the first P frames are:

- Temporal features: Frame difference in RGB-D feature space or ($Temp\_Val_1$) from Eq. (3.2), denoted as T_{RGB} and T_D, respectively.
- Color features: The RGB-D values present in the object model, denoted as RGB_V, D_V.
- Spatial features: The probable location of the object region in the current frame, (generally taken as 1.5 times of the difference region between successive frames), denoted as Sp_L.

Table 3.1: Rule generation for object–background separation in pixel level.

U	T_{RGB}	T_D	RGB_V	D_V	Sp_L	$Decision$
1	0	0	Ou	Ou	Ou	B
2	0	0	W	W	Ou	B
3	0	0	W	W	Ou	O
4	1	0	W	W	W	B
5	1	1	Ou	Ou	W	O
6	1	1	W	Ou	W	O
7	1	0	Ou	W	W	B
8	1	1	Ou	Ou	Ou	O
9	1	1	Ou	Ou	Ou	B
10	0	1	W	Ou	W	O
11	1	1	W	W	Ou	O
12	1	1	W	W	W	O

The rule-base formed with these conditional features is described in the following section.

3.4.2 *Rule-base in Pixel Level*

The decision table containing the generated rules is shown in Table 3.1 (Chakraborty and Pal, 2015). The first two conditional attributes (T_{RGB} and T_D) have the cardinality of 2 (0 and 1) which signifies change (1) or not change (0), and the last three conditional attributes (RGB_V, D_V and Sp_L) are also of cardinality two indicating within the model (W) and outside (Ou).

Twelve various observations representing different cases are shown in Table 3.1. The cases are as follows:

(1) An ideal background pixel.
(2) A background pixel with same features as that of an object.
(3) An object pixel which moves slower than estimated.
(4) A similar colored background pixel that was inside the object in the previous frame.
(5) An object pixel with changing depth (movement towards or away from camera) along with its movement.

(6) A moving object pixel within the object area, i.e., object-to-object pixel.

(7) A background pixel within region of interest with changed RGB value, i.e., an object-to-background pixel.

(8) An object pixel which starts to move from the current frame onwards.

(9) A noisy background pixel.

(10) An object pixel moving in similar RGB-D background.

(11) An object pixel which moves faster than estimated.

(12) An ideal object pixel or background to object pixel.

It can be noticed from Table 3.1 that *Rules* $2, 3$ and $8, 9$ are inconsistent. That means, the decision taken according to the rules may not be true. The remaining seven rules are true. So, the dependency between the condition and decision attributes is: $8/12$.

To eliminate the superfluous attributes, the core of the conditional attributes is computed. It is seen that elimination of the feature T_{RGB} does not make the rule-base indiscernible or reduce the dependency. Hence, T_{RGB} is not a core. Whereas rules 4 and 12 become indiscernible by the elimination of T_D. Individual elimination of RGB_V or D_V does not affect the dependency, but together it affects. Hence, these two attributes can be in the core. Sp_L is also a core. That means, out of the five reducts (features), four are cores. Based on the concept of core, the rules that lead to maximum correct decisions are defined as follows:

(1) If $T_D = 1$ and
$RGB_V = W$ or $D_V = W$ and $Sp_L = W$, then decision $= O$.

(2) If $T_D = 0$ and
$RGB_V = Ou$ or $D_V = Ou$ and $Sp_L = Ou$, then decision $= B$.

This is a faster process of tracking. This method is well suited for ideal object–background separation. But, when there are some other challenges or the cases like 2, 3, 8, 9 occur, this method fails. In search of having more accurate process or to increase the dependency between the condition and decision attributes, the concept of rule-base with 3-D neighborhood granules was introduced in (Pal and Chakraborty, 2017) keeping all of the input information the same. The method is explained in the following section.

3.4.3 *Rule-base in Granular Level*

Here the conditional features considered to form the rule-base were:

- Spatio-Temporal features in RGB-D space: Values of frame difference according to Eq. 3.2.
- Color features in RGB-D space: RGB-D values present in Q .

Therefore, the conditional features in granular level came down to: (i) 3-D spatio-temporal granules ($\aleph_{sp-tmp}$) according to Eq. (3.4) over the region Q, (ii) color/RGB-D granules ($\aleph_{rgb}$) and ($\aleph_D$) formed over the color values of Q.

The corresponding new rule-base is then generated by incorporating the 3-D spatio-temporal granules and color granules. This granulated rough rule-base is shown in Table 3.2. Here there are four attributes for each feature: (i) belongs (Be), (ii) does not belong (NB), (iii) partially belongs (PB) and (iv) contained in (CC). Rules 1 and 12 represent the ideal cases, and the remaining ambiguous.

It can be seen from Table 3.2 that, rules 2, 3 are no more inconsistent. That means, the pixel level features which are crisp in nature could not model the differences between cases 2 and 3, as expected. On the other hand, the granular level features which have overlapping boundaries can do it; thereby demonstrating its appropriateness which is natural too. The rule-base, thus generated, is able to classify

Table 3.2:   Rule-base with 3-D neighborhood granules.

U	$\aleph_{sp-tmp}$	$\aleph_{RGB}$	$\aleph_D$	*Decision*
$\aleph_1$	NB	NB	NB	B
$\aleph_2$	NB	Be	Be	B
$\aleph_3$	PB	Be	Be	O
$\aleph_4$	PB	Be	PB	B
$\aleph_5$	Be	Be	PB	O
$\aleph_6$	CC	Be	Be	O
$\aleph_7$	NB	NB	Be	B
$\aleph_8$	Be	NB	NB	B
$\aleph_9$	Be	NB	NB	O
$\aleph_{10}$	PB	PB	CC	O
$\aleph_{11}$	Be	Be	Be	O
$\aleph_{12}$	Be	Be	Be	O

all the cases, expect the presence of noise. The dependency between condition and decision attributes is now 10/12 as compared to 8/12 in Table 3.1.

One may note that the rules that are not relevant with respect to the problem of object tracking have been eliminated during the formation of this rule-base. These aforesaid 12 rules characterize the situations that occur in a sequence often. It was proven that the set of these rules gives more than 90% coverage. In this way, 12 most relevant rules were considered out of $4^3 = 64$ probable combinations. For example, the rule [*CC NB NB*] does not have any significance in this scenario, as no granule contained by the spatio-temporal region can have totally different color granules than that of its estimated object region.

The rule-base (Table 3.2) was formed using the initial object/background model and the observation over P number of initial frames. The method of acquiring the conditional features based on the said observations is described in the following section.

3.5 Adaptive Rule-base: A Granular Flow Graph Approach

3.5.1 *Information Flow Graph*

Information flow graph was introduced (Pawlak, 2005) to model the information flow in a system. A flow graph is a directed, acyclic and finite graph represented as $G = (N, \mathcal{B}, \varphi)$. N represents the total number of nodes, $\mathcal{B} \subseteq N \times N$ represents a set of directed branches, $\varphi : \mathcal{B} \rightarrow R^+$ is a flow function and R^+ is the set of non-negative reals. Each node of a flow graph represents an attribute of the information system. The input and output represented as n_i and n_o of a node n are the sets: $n_i = m \in N : (m, n) \in \mathcal{B}$ and $n_o = m \in N : (n, m) \in \mathcal{B}$.

The inflow and outflow of that node n are computed as follows:

$$\varphi_+(n) = \sum_{m \in n_i} \varphi(m, n) \tag{3.8a}$$

$$\varphi_-(n) = \sum_{m \in n_o} \varphi(n, m) \tag{3.8b}$$

where $\varphi(m, n)$ is the flow function characterizing the flow from node m to node n. The inflow and outflow of an internal node of a graph are supposed to be the same, and so of the graph G. Let $\varphi(G)$ represent the through-flow of the graph G.

The normalized flow graph is represented as $G = (N, \mathcal{B}, \sigma)$ where N and $\mathcal{B}$ are the same as before, and

$$\sigma(m, n) = \frac{\varphi(m, n)}{\varphi(G)}. \tag{3.9}$$

The certainty (cer) and coverage (cov) are two major factors associated with every branch of a flow graph. These are defined for the branch (m, n) as

$$cer(m, n) = \frac{\sigma(m, n)}{\sigma(m)} \tag{3.10a}$$

$$cov(m, n) = \frac{\sigma(m, n)}{\sigma(n)}. \tag{3.10b}$$

$\sigma(m)$ and $\sigma(n)$ are the normalized weights of the nodes m and n, respectively, and $\sigma(m, n)$ is the normalized weight of the branch (m, n).

3.5.2 *Flow Graph in Video: Relevance and Issues*

There are several scenarios in a video sequence when the initially defined rule-base can not give much effective results, rather gives many false positive or false negative outputs. That is, the importance of a certain rule in a system may increase or decrease with time. For example, new moving object(s) may appear into a sequence, or moving object(s) may get stopped or disappear from the sequence. The new appearance of object(s) needs new sets of conditional attributes, whereas disappearance of object(s) needs deletion of the respective sets of the conditional attributes. These can be detected with proper updation of the rule-base. Flow graph shows the relationship between the attributes (nodes) and the relevance of the rules (branches) generated from those attributes. Therefore, flow graph has the ability to detect the deviation in the value of a particular branch. The updation of only the attributes associated with the branch can be an effective technique for rule-base adaptation. In this way the sudden or unpredictable changes in a video can be detected and the corresponding set of rules can be updated. It may be noted that the updation is mostly required when there are multiple moving objects in a sequence.

As discussed earlier, all of the conditional features can not remain the same over time in a video sequence, rather these should change and get updated depending on the change in shape, alignment, color of the object(s). The updation is also required with the appearance of new object(s) or if a still object starts to move. The updation of all the conditional features in every frame is quite time-consuming and hence less convenient. Therefore, an intelligent rule-adaptation process is required. Let us describe such an intelligent system. The underlying idea behind this process is that the attributes should get updated when required instead of in every frame. Here, all of the conditional attributes do not get updated simultaneously, rather it is done according to the requirement. This is decided with the help of the flow graph.

In the said approach, the relevance of a certain rule or attribute in an information system is mapped with the help of flow graph. When some deviation in the values of any branch/node is detected, the updation of only the attributes associated with those branches is expected to give the desired results. The flow graph that is modeled here is on granular level and the information flow is designed with the neighborhood granules. Therefore it is named as 'granular flow graph'.

In this process, a training flow graph is designed first based on the labeled information. The foreground regions of the training frame are the input to the training flow graph. The initial values of its branches and nodes are set to those, as obtained from the rule-base in Table 3.2.

An example of training flow graph according to training data set of $2b$ sequence (described in Section 3.4) is shown in Fig. 3.3. Initially there are 300 object granules and 700 background granules represented as $\varphi(O) = 300$ and $\varphi(B) = 700$, respectively. The branch $\varphi(O, NB_T) = 20$ indicates that only 20 granules out of those 300 object granules do not belong to the set of the temporal values (NB_T). The values assigned to the rest of the branches represent the similar characteristics. The output shows the correctness in the classification. For example, $D1_{+ve} = 910$ means that 910 granules out of the 1000 input granules are correctly classified. The performance of the rule-base is measured over each input frame with the help of training flow graph.

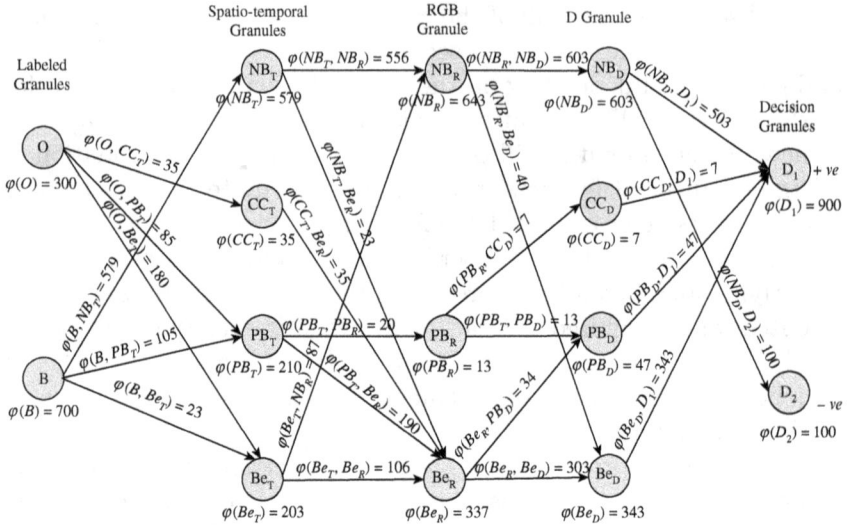

Fig. 3.3:   Example flow graph model according to Table 3.2.

The test flow graph is generated afterwards over each input frame (current frame). The test flow graph is compared with the training flow graph with respect to node and branch parameters. The test flow graph is designed with the assumption that all of the spatio-temporal neighborhood granules with higher gray level values (i.e., expected moving object) are the foreground granules. Hence, these granules are given as the input to the test flow graph. The entire test flow graph, corresponding to an input frame, does not get formed at a time, rather the values are assigned to every node and branch after checking their deviations from the corresponding training values. If the deviation in a certain node or branch is more than the expected, the associated sets of attributes get updated. The updation is done based on the spatio-temporal information in the current frame and this is applied from the next frame onwards.

In case of videos, the number of foreground-background granules can not remain the same over the sequence, rather these are expected to change from frame-to-frame. The comparison between two flow graphs (training and testing) for updation of the latter is fair as long as the input dataset distribution among the classes for those graphs is of the same ratio. But, the ratio cannot remain the same if the

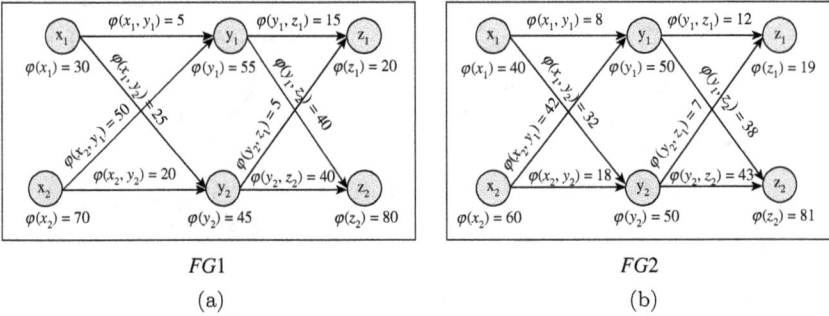

FG1 FG2

(a) (b)

Fig. 3.4: Two example graphs (a) training flow graph and (b) test flow graph.

distribution of datasets among the classes gets changed. In Fig. 3.4, such a case is shown with two simple example flow graphs. There are two input classes with data distributions 3:7 and 4:6. Let $FG1$ be the training graph and $FG2$ be the test graph. The overall output accuracies for both are almost the same. However, if the normalized weights in every branch and node of the two graphs are compared, then a large deviation can be found. It will lead to false decisions. Therefore, comparison of these two flow graphs is unfair for updation of the test flow graph $FG2$ and updation of the rule-base for these cases is not justifiable.

The concepts of expected flow graph and mutual dependency between flow graphs were introduced in Pal and Chakraborty (2017) to deal with the aforesaid situation and to make the flow graph applicable to verification tasks. These are described in the following sections.

3.5.3 *Expected Flow Graph*

Expected flow graph characterizes how the training flow graph is expected to be modeled if the data distributions among the classes get changed. Two simple flow graphs are shown in Fig. 3.5 for elaboration.

The graph $G1$ is the normalized training graph with two input classes with data distribution of $a : b$. $G1'$ is the expected model of $G1$ if the data distribution gets changed to $\alpha : \beta$. The respective values of the branch parameters, viz, m, n, o, p get changed to m', n', o', p'

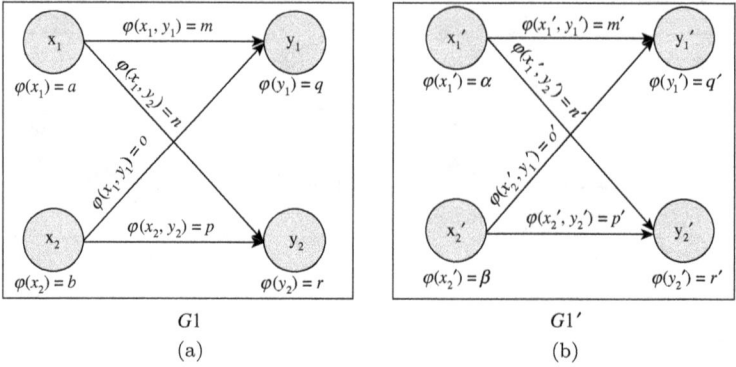

G1
(a)

G1'
(b)

Fig. 3.5: Example: Two similar flow graphs with different data distributions.

as follows:

$$m' = \frac{m\alpha}{a} = \frac{m}{a} \times \alpha = cer(x_1, y_1) \times \alpha \qquad (3.11a)$$

$$n' = \frac{n\alpha}{a} = \frac{n}{a} \times \alpha = cer(x_1, y_2) \times \alpha \qquad (3.11b)$$

$$o' = \frac{o\beta}{b} = \frac{o}{b} \times \beta = cer(x_2, y_1) \times \beta \qquad (3.11c)$$

$$p' = \frac{p\beta}{b} = \frac{p}{b} \times \beta = cer(x_2, y_2) \times \beta \qquad (3.11d)$$

$cer(.)$ is defined as in Eq. (3.10).

The expected values of the nodes in G'_1 are

$$q' = m' + o' = cer(x_1, y_1) \times \alpha + cer(x_2, y_1) \times \beta \qquad (3.12a)$$

$$r' = n' + p' = cer(x_1, y_2) \times \alpha + cer(x_2, y_2) \times \beta \qquad (3.12b)$$

For example, the expected normalized flow graph of $FG1$ in Fig. 3.4 with data distribution changed to $3:7$ from $4:6$ (which is the case of $FG2$) is shown in Fig. 3.6. Let it be named as $FG1'$. Here, the values of the nodes and branches are computed as per Eqs. (3.12) and (3.11), respectively.

3.5.4 *Mutual Dependency Between Flow Graphs*

The mutual dependency between two flow graphs represents how reliable the decision of a graph would be if it was taken by the other

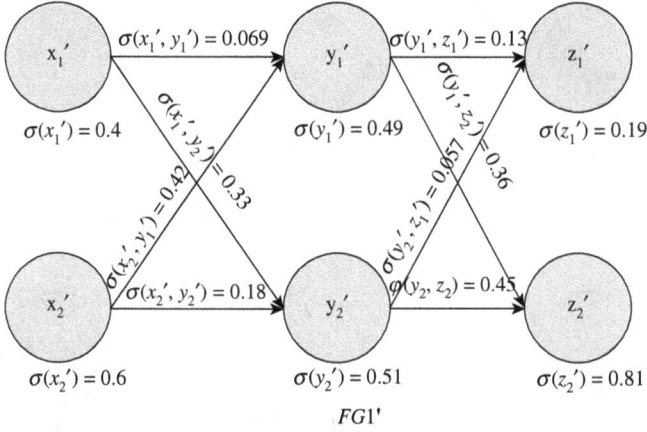

Fig. 3.6: FG1': Expected flow graph of $FG1$ in Figure 3.4 with different distributions.

one. That is, if a decision is reliable with $Graph_1$, how reliable could it be with $Graph_2$? This is computed by dividing two normalized flow graphs with similar nodes and branches, but the weights can be different.

Let $NFG_1 = (N_1, \mathcal{B}_1, \sigma_1)$ and $NFG_2 = (N_2, \mathcal{B}_2, \sigma_2)$ be two normalized flow graphs. The mutual dependency between them can be calculated if and only if:

- $N_1 \equiv N_2$
- $\mathcal{B}_1 \equiv \mathcal{B}_2$
- $\sigma_1 \neq \sigma_2$
- $\sigma(N_{1i_1}) : \sigma(N_{1i_2}) : \ldots : \sigma(N_{1i_C}) = \sigma(N_{2i_1}) : \sigma(N_{2i_2}) : \ldots : \sigma(N_{2i_C}),$

where N_{1i} and N_{2i} are the input nodes of NFG_1 and NFG_2, respectively. C is the total number of input classes and the symbol '$\equiv$' represents equivalence.

The mutual dependency between two of them is denoted by μ and computed as

$$\mu \equiv \left\{ \frac{NFG1}{NFG2} \right\}. \tag{3.13}$$

The mutual dependency between the respective nodes and branches of the two flow graphs under consideration is computed

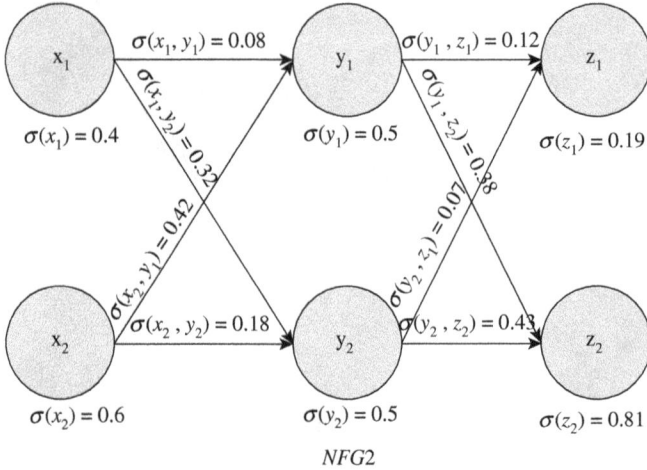

NFG2

Fig. 3.7:   NFG2: Normalized flow graph of FG2.

with Eq. (3.13). As the initial two elements of the set μ are supposed to be unity (1) according to the definition, only the value of $\frac{\sigma_1}{\sigma_2}$ is considered during comparison.

If the two graphs have same weights, all the nodes and branches of their mutual dependency graph will have the value unity. That is, same decision-making can be performed from both of the graphs. The more the deviation from one (unity), the lesser the reliability of the respective node(s) or branch(es). The normalized flow graph of $FG2$, denoted as $NFG2$, is shown in Fig. 3.7.

It can be noticed that the two graphs $NFG2$ and $FG1'$ (Fig. 3.6) satisfy the aforesaid conditions for computing the mutual dependency. The computed mutual dependency between them is shown in the graph in Fig. 3.8. It can be observed that the values of the branches (x_1, y_1), (x_1, y_2), (y_2, z_1) and (y_2, z_2) deviate more from the ideal value of one (1). Hence the nodes (y_1 and y_2) representing the conditional attributes and the corresponding rules which establish the above connections need updation. The threshold which detects when the updation is required based on the aforesaid derivation is application-dependent.

These two concepts, expected flow graph and mutual dependency between flow graphs, were used in the decision-making process of

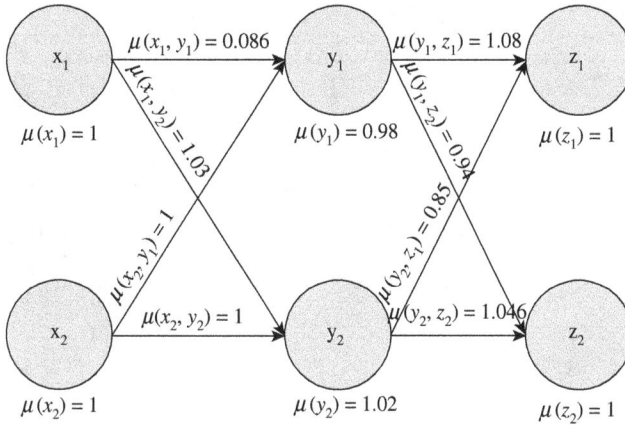

Fig. 3.8: Mutual dependency between FG1' and NFG2.

NRBFG method for tracking. The expected flow graph and the training flow graph were designed according to the data distribution in the current frame. The mutual dependency between the normalized flow graph of current (input) frame and expected training flow graph was then measured. For the values where the deviations are much higher or lower than unity, the respective set of attributes got updated. The steps involved in flow graph-based adaptation are described here in Algorithm 1.

3.6 Experimental Results

Some experimental results along with comparison (Chakraborty and Pal, 2015; Pal and Chakraborty, 2017) are given here in the present section to demonstrate the effectiveness of NRBFG method in tracking multiple objects. The experiment was carried out with different types of video sequences and more than 2500 frames in total. Experimental results establish the effectiveness of the following:

- Rough rule-base for unsupervised video tracking, both in granular and pixel levels.
- Neighborhood granular rough rule-base in tracking over crisp granular and pixel level rule-bases.
- Adaptive rule-base with rough flow graph.

Algorithm 1 Flow-graph Based Adaptation

INPUT: Training flow graph $FG1$ and test flow graph $FG2$

OUTPUT: Set of features $\mathcal{F}$ that need to be updated

Set $\mathcal{F} =$

1: Generate expected flow graph $FG1'$ of $FG1$.

2: Compute normalized flow graph $NFG2$ of $FG2$.

3: Evaluate the mutual dependency graph (μFG) between $FG1'$ and $NFG2$.

4:

if $\mu(z_1) = \mu(z_2) = 1$ **then**

    No updation is required.

else

    Detect the branches $(b_1, ..., b_n \in \mathcal{B}_1)$ where $|\mu(b) - 1| > \delta$ and their associate nodes $n1, ...n_n \in N_1$. Set $\mathcal{F} = \{n1, ...n_n\}$.

end if

5: Update the range of the features in $\mathcal{F}$ associated to the nodes by considering the information from previous P frames.

- Visual and quantitative performance of NRBFG method with respect to other popular methods.

3.6.1 *Results of Tracking*

Results of NRBFG method of tracking are shown here on several RGB-D video sequences as acquired by Kinect sensor and IR sensor. These were obtained from CGD-2011 (ChaLearn, 2011) and Thermal and Visible Imagery (Davis and Sharma, 2007). The frame size in each sequence is of 240×320 pixels. In those videos there are many sequences with different types of movements of human hands which are sensed by kinect sensor, and several surveillance scenarios with movements of people obtained from IR sensor. The results over a few such types of sequences (viz, M_2, M_4, M_9 from CGD-2011, and $2b$ and $5b$ from Thermal and visible Imagery) are shown here, as examples, for $P = 6$ (the number of previous frames). The initial value of Thr_t (in Eq. 3.4) was assigned as $0.2 \times max(\tau_{med})$, with an assumption that at least 20% pixels of a frame will be within the moving object. However, if the number of 3-D granules gets changed from frame-to-frame and only a few granules remain same over all the frames and others vary, then Thr_t will get decreased by 0.5 times.

It will be increased by 1.5 times in the reverse scenario, i.e., when there will be very few moving granules detected in each frame. Thr_c (Eq. 3.5) was initially assigned as 5, according to Weber's Law.

3.6.1.1 *Comparisons Between Neighborhood Granular Level and Pixel Level Methods*

The comparative study refers between the neighborhood granular level (NGrRB) (as in Section 3.4.3) and a pixel level (PRB) (as in Section 3.4.2) rule-base methods without any adaptation. The results are shown in Table 3.3. The metrices over which the comparison is made are (i) accuracy which is measured based on the distance between the centroids (CD) of the ground-truth and the obtained foreground segment of the respective frames and (ii) time (average CPU processing time (in seconds) for each frame). The ground-truths that were used were manually annotated in every frame. If the CD is less, it means more accuracy. The experimental results over two video sequences are shown here. The sequence M_2 contains a hand movement scenario and the sequence $5b$ contains a surveillance scenario.

It can be seen from Table 3.3 that the pixel level method (PRB) is a bit faster but is less accurate. These results lead to the theoretical conclusion that the PRB provides less indiscernibility than that of NGrRB, thereby producing inferior performance.

Note that the granules considered in NGrRB are overlapping granules with the attribute set of higher cardinality and hence the computation with these granules makes the process little slower compared to PRB. But, it does not contradict the preliminary assumption that granulation makes a process faster. Crisp (i.e., non-overlapping) granulation with same cardinality that of PRB obviously makes the process faster, but at the cost of accuracy.

Table 3.3: Computation time and accuracy of NGrRB and PRB.

Sequence	Method	Avg. CPU Time/Frames	Max Time	Min Time	CD
M_2	$NGrRB$	0.381	0.631	0.192	3.24
M_2	PRB	0.331	0.394	0.213	6.11
$5b$	$NGrRB$	0.228	0.366	0.163	3.52
$5b$	PRB	0.186	0.267	0.155	7.16

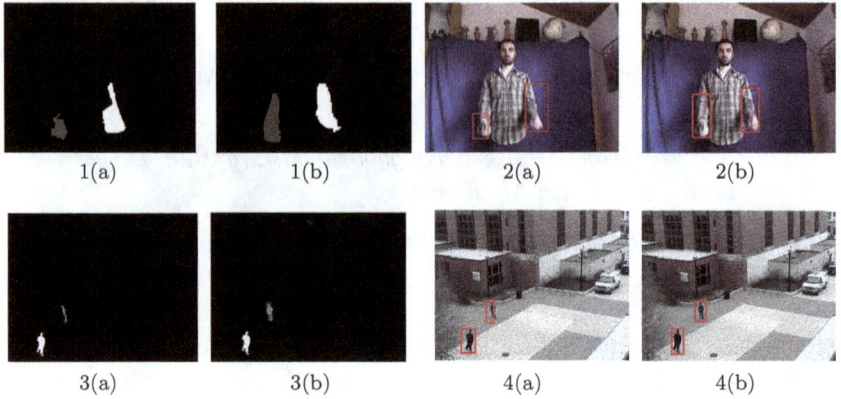

1(a)   1(b)   2(a)   2(b)

3(a)   3(b)   4(a)   4(b)

Fig. 3.9: Visual comparison between the techniques (a) PRB and (b) NGrRB over Frame no. 12 of M_2 sequence (1, 2) and Frame no. 415 from 5b sequence (3, 4).

The two visual results of Frame number 12 from M_2 sequence and Frame no. 425 from 5b sequence, as shown in Fig. 3.9, depict that *PRB* cannot detect the total object when it is moving slower than expected, and treats some parts as noise. These were the cases characterized by rules 2 and 3 (Section 3.4.2), and this indiscernibility could not be overcome. In the first scenario (Figs. 3.9.1 and 3.9.2), some parts of the hands are moving slower, whereas in the second scenario (Figs. 3.9.3 and 3.9.4) one object (human) is moving slower. Neither both the hands, nor both the persons can totally be detected as the foreground segment by PRB (see Figs. 3.9.1(a) and 3.9.3(a)), i.e., misclassification occurs. The detection and tracking is more accurate with NGrRB (see Figs. 3.9.1(b), 3.9.3(b)), as expected.

The results shown in Fig. 3.9 demonstrate the effectiveness of the neighborhood granular rule-base over pixel level one. The effect of rule-base updation with flow graph is shown in the following section.

3.6.1.2 Comparisons Between with Flow Graph and Without Flow Graph

Without flow graph, there may be two actions, the rule-base either will not get updated (NGrRB) or the entire rule-base will get updated in every frame (RU). Here, the comparative studies of these instances with the flow graph-based method (NRBFG) are shown in Table 3.4.

Table 3.4: Computation time and accuracy of NGrRB, RU and NRBFG.

Sequence	Method	Time	Coverage(%)
M_9	$NGrRB$	0.222	72
M_9	RU	0.836	97
M_9	$NRBFG$	0.335	95
M_4	$NGrRB$	0.308	69
M_4	RU	0.798	96
M_4	$NRBFG$	0.369	94

The time (avg CPU time/frame in sec.) and rule-base coverage are the metrices considered in this table. The two datasets over which the results are shown in Fig. 3.10 represent two types of hand movements where NGrRB fails.

In case of sequence M_9, the left hand of the person was moving initially, and the right hand starts its movement from 30th frame onwards. As expected, the movement of the right hand cannot be classified by NGrRB (Figs. 3.10.1(a) and 3.10.1(b)) due to the inconsistency between the rules 8 and 9 as in Table 3.2 which results in less coverage (as shown in Table 3.4). However, it is successfully tracked by NRBFG (see Figs. 3.10.1(b) and 3.10.2(b)). The similar situation occurs in case of M_4 sequence, where the right hand was moving initially and it gets stopped at 17th frame, and the left hand starts its movement. Nothing can be tracked by NGrRB (see Figs. 3.10.3(a) and 3.10.4(a)) in this scenario, whereas the moving left hand can be successfully tracked by NRBFG (see Figs. 3.10.3(b) and 3.10.4(b)). It is evident from Table 3.4 that RU gives good results but is more time-consuming as expected, whereas NRBFG keeps a good balance between the time and accuracy.

3.6.2 *Comparisons*

Here we provide the comparative performance of NRBFG with five recent tracking methods, namely, sequential partial filtering on graphs (SPG)(Pan and Schonfeld, 2011), PLS-tracking (Wang *et al.*, 2012), continuous energy minimization-based multiple target tracking (CEMT) (Milan *et al.*, 2014), sparse representation-based local appearance learning for tracking (LLAS) (Bai *et al.*, 2015), and

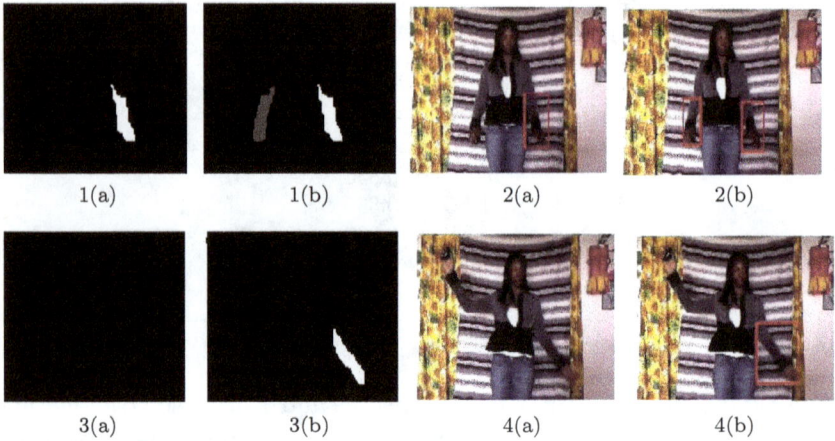

Fig. 3.10: Visual comparison between the techniques (a) NGrRB and (b) NRBFG over Frame no. 30 of M_9 sequence (1, 2) and Frame no. 17 from M_4 sequence (3, 4).

Grassmann Manifolds-based online tracking method GMOT (Khan and Gu, 2013). The SPG method is useful for surveillance tracking, PLS is useful for partial occlusion handling, CEMT and LLAS are effective both in case of multiple target tracking and overlapping handling, and GMOT is effective in handling changing shapes of object(s). All of them are partially supervised methods, i.e., initial manual labeling is required. The visual and quantitative comparative results for four sequences are given in Fig. 3.11 and Table 3.5. The characteristics of all the four sequences are as follows. The sequences M_9 and M_7 represent two hand movement videos sensed by kinect sensor. In M_9, the left hand of the person was moving initially with changing shapes and sizes and the right hand starts to move from 30th frame onwards (see Fig. 3.11.1). In M_7 sequence, both the hands of the person are moving, getting overlapped, and partially occluded from each other ((see Fig. 3.11.2)). The sequences $5b$ and $2b$ are two surveillance scenarios obtained from IR sensor. In $5b$, two persons are moving with different speeds (see Fig. 3.11.3). In the sequence $2b$, six persons are moving in different directions and speeds, one of them is moving in similar colored background and is hardly separable in RGB feature space (see Fig. 3.11.4). In Fig. 3.11, the red tracker shows the results obtained by NRBFG, the green tracker shows the results of PLS, the blue tracker shows the results of SPG, the yellow

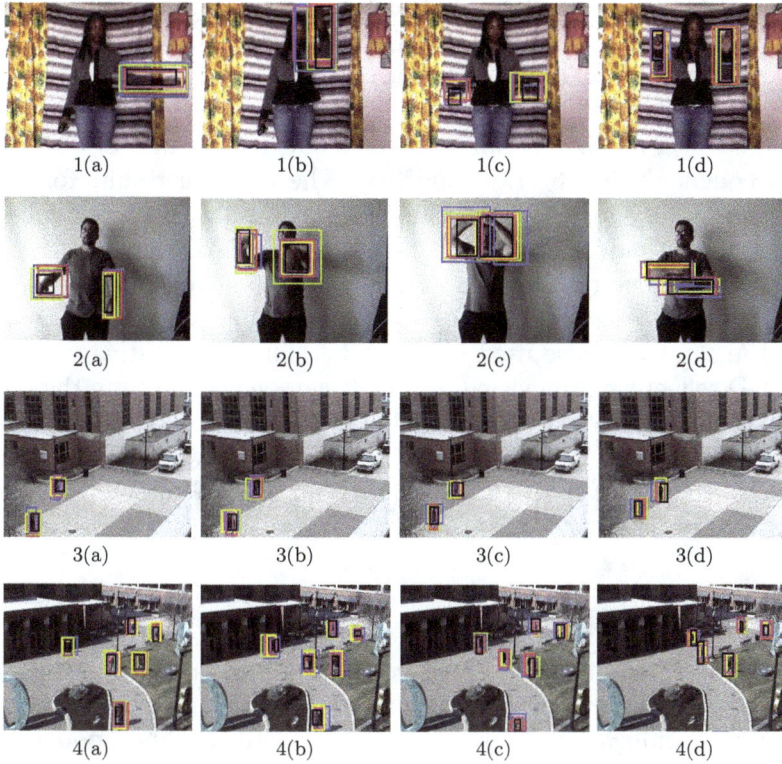

1(a)   1(b)   1(c)   1(d)

2(a)   2(b)   2(c)   2(d)

3(a)   3(b)   3(c)   3(d)

4(a)   4(b)   4(c)   4(d)

Fig. 3.11: Tracking results for Frame nos. (1) 13, 22, 32, 39 from M_9 sequence, (2) 11, 23, 30, 42 from M_7 sequence, (3) 370, 440, 471, 510 from $5b$ sequence and (4) 44, 72, 132, 250 from 2b sequence.

Table 3.5: Computation time and accuracy of NRBFG, PLS, SPG, CEMT, LLAS and GMOT.

Sequence	Metric	NRBFG	PLS	SPG	CEMT	LLAS	GMOT
M_9	CD	4.78	31.42	7.06	4.23	4.42	4.75
M_9	Time	0.325	0.275	0.390	0.352	0.348	0.365
M_7	CD	5.54	7.21	6.86	4.94	4.88	6.21
M_7	Time	0.258	0.323	0.356	0.283	0.265	0.289
$5b$	CD	4.23	6.32	6.22	3.67	3.92	4.11
$5b$	Time	0.308	0.341	0.395	0.325	0.371	0.398
2b	CD	5.52	7.02	7.42	4.25	4.12	6.22
2b	Time	0.386	0.441	0.495	0.422	0.451	0.462

tracker shows the results obtained by CEMT, the magenta tracker shows the results obtained by LLAS, and the black tracker shows the results of GMOT. It can be noticed that all of the sequences are successfully tracked by NRBFG. The overlapping scenarios are better handled by PLS, but it fails to detect new appearance of object in the sequence (Figs. 3.11.(c) and (d)). The tracking results for SPG are satisfactory for all of the four sequences, but the tracker fails to cover all the moving objects, whereas both CEMT and LLAS give very satisfactory results even with multiple moving elements.

The matrices considered in Table 3.5 for quantitative comparison are the same as those used in Table 3.3. The quantitative values of the CD reflect well the visual results. It can be noticed from the table that the computation time is the lowest for NRBFG with accuracy better than PLS, SPG and GMOT, but slightly worse than CEMT and LLAS. However, NRBFG requires no manual interactions, i.e., it is unsupervised.

3.7 Conclusions and Discussions

The problem of adaptive rule generation for unsupervised video tracking in granular computing frame work has been addressed in this chapter. Significance of rough flow graph and neighborhood rough sets is explained. The study demonstrates how the concept of rough rule-base can be used over 3-D granulated space for unsupervised tracking or tracking without manual interactions, how to formulate a method of updating the rough rule-base using rough flow graph, and introduces few new features like expected flow graph and mutual dependency between flow graphs. This also includes a method of forming 3-D natural granules, and formulating quantitative indices for tracking based on these granules and IR sensed data.

Rough flow graph provides a method of intelligent adaptation of the rule-base, and the method NRBFG based on this is proven to be very effective both in terms of accuracy (as compared to NGrRB) and time (as compared to RU), as expected. Two added features, namely, expected flow graph and mutual dependency between flow graphs, defined over the flow graph, make it applicable for testing and validation of the rules. NRBFG provides the best tracking results compared to the other two non-adaptive methods (PRB and NGrRB), however, it takes maximum time among them. Further, incorporation of flow graph for rule-base adaptation enables detection of newly appeared objects in a sequence, which is treated as noise in the non-adaptive method NGrRB.

The comparative performance among NRBFG and five recent partially supervised methods shows that PLS has superior ability in handling occlusion, GMOT can handle the variations in object shapes better, SPG can detect the unpredictable changes in videos more accurately, whereas CEMT and LLAS are effective in handling both of the cases. On the other hand, all these problems can be successfully handled by the unsupervised NRBFG with a good balance between time and accuracy.

Granular level computation using 3-D spatio-temporal neighborhood granules is superior to pixel level computation in increasing the discrimination ability of the rule-base as well as in reducing the computation time. Indices based on these granules (without ground-truth information) are seen to reflect well the quality of unsupervised tracking.

So far, we have demonstrated in Chapters 2 and 3 the effectiveness of rough set theoretic granular computing in the tasks of partially supervised tracking and unsupervised tracking. The major issues in the task of tracking involve occlusion of moving object(s) to the background and overlapping among the moving objects. The effectiveness of rough set theories in handling such tasks and producing accurate approximations is given in detail in Chapter 4.

Bibliography

Bai, T., Li, Y.-F. and Zhou, X. (2015). Learning local appearances with sparse representation for robust and fast visual tracking, *IEEE Trans. Cyberns.* **45**(4): 663–675.

Chakraborty, D. B. and Pal, S. K. (2015). Neighborhood granules and rough rule base in tracking, *Nat. Comput. Springer (Special Issue on Pattern Recognition and Mining)*, **15**(3): 359–370.

ChaLearn (2011). *ChaLearn Gesture Dataset (CGD 2011)*, California.

Chi, Z., Li, H., Lu, H. and Yang, M.-H. (2017). Dual deep network for visual tracking, *IEEE Trans. Image Proc.* **26**(4): 2005–2015.

Cucchiara, R., Grana, C., Piccardi, M. and Prati, A. (2003). Detecting moving objects, ghosts and shadows in video streams, *IEEE Trans. PAMI* **25**: 1337–1342.

Davis, J. and Sharma, V. (2007). Background-subtraction using contour-based fusion of thermal and visible imagery, *Comput. Vis. Image Underst.* **106**: 162–182.

Elgammal, A., Duraiswami, R., Harwood, D. and Anddavis, L. (2002). Background and foreground modeling using nonparametric kernel density estimation for visual surveillance, *Proc. IEEE 90*, Vol. 7, pp. 1151–1163.

Jain, R. and Nagel, H. (1979). On the analysis of accumulative difference pictures from image sequences of real world scenes, *IEEE Trans. PAMI* **1**(2): 206–214.

Khan, Z. and Gu, I.-H. (2013). Nonlinear dynamic model for visual object tracking on grassmann manifolds with partial occlusion handling, *IEEE Trans. Cyberns.* **43**(6): 2005–2019.

Lisowski, K. and Czyzewski, A. (2014). Modelling object behaviour in a video surveillance system using pawlak's flowgraph, *7th International Conference, MCSS 2014*, Poland, pp. 122–136.

Maddalena, L. and Petrosino, A. (2008). A self-organizing approach to background subtraction for visual surveillance applications, *IEEE Trans. Image Process.* **17**: 1168–1177.

Milan, A., Roth, S. and Schindler, K. (2014). Continuous energy minimization for multitarget tracking, *IEEE Trans. PAMI* **36**(1): 58–72.

Pal, S. K. and Chakraborty, D. B. (2017). Granular flow graph, adaptive rough rule generation and tracking, *IEEE Trans. Cyberns.* **47**(12): 4096–4107.

Pan, P. and Schonfeld, D. (2011). Video tracking based on sequential particle filtering on graphs, *IEEE Trans. on Image Proc.* **20**(6): 1641–1651.

Pawlak, Z. (2005). *Flow Graphs and Data Mining*, Springer-Verlag, Berlin, Heidelberg.

Rittsher, J., Kato, J., Joga, S. and Blake, A. (2000). A probabilistic background model for tracking, *European Conf. on Comput. Vis.* **2**: 336–350.

Stauffer, C. and Grimson, W. E. L. (2000). Learning patterns of activity using real-time tracking, *IEEE Trans. PAMI* **22**: 747–757.

Sun, S., Akhtar, N., Song, H., Mian, A. and Shah, M. (2017). Deep affinity network for multiple object tracking, *IEEE Trans. PAMI* **13**(9): 1–15.

Voigtlaender, P., Krause, M., Osep, A., Luiten, J., Sekar, B. B. G., Geiger, A. and Leibe, B. (2019). Mots: Multi-object tracking and segmentation, *IEEE Conference of Computer Vision and Pattern Recognition (CVPR)*.

Wang, Q., Chen, F., Xu, W. and Yang, M.-H. (2012). Object tracking via partial least squares analysis, *IEEE Trans. Image Proc.* **21**(10): 4454–4465.

Wang, Q., Zhang, L., Bertinetto, L., Hu, W. and Torr, P. H. (2019). Fast online object tracking and segmentation: A unifying approach, *IEEE Conference on Computer Vision and Pattern Recognition (CVPR)*.

Wren, C., Azarbayejani, A. and Pentland, A. (1997). On the analysis of accumulative difference pictures from image sequences of real world scenes, *IEEE Trans. PAMI* **19**(7): 780–785.

Chapter 4

Unsupervised Occlusion Handling

4.1 Introduction

Tracking object(s) from videos when they are partially overlapped or fully occluded is another major challenge in the task of video tracking. The focus of this chapter is mainly on showing the effectiveness of rough sets in handling these difficult issues. As stated before, the task of tracking moving object(s) from video sequences becomes difficult when there exist several uncertainties arising from, e.g., changes in shapes/sizes of moving object(s), changes in motion of the object(s), and changes in the number of object(s), occurrence of object-to-object overlapping, and occurrence of total/partial occlusion to background. These problems were addressed and several approaches were formulated using statistical modeling, local features extraction etc. (Hu *et al.*, 2015; Park *et al.*, 2015; Wang *et al.*, 2015; Milan *et al.*, 2014; Pernici and Bimbo, 2014). But, the initial object(s) need to be labeled manually in all those approaches. Besides, those methods may fail to estimate the locations of the object(s) during their total occlusion to background. The neighborhood rough filter-based tracking (NRFT) method (Chakraborty and Pal, 2018) described in the chapter did not need any initial manual interaction and proved to be effective in handling total occlusion. The merits of neighborhood rough set (NRS) theories were exploited in NRFT to model the uncertainties which are present in the tasks of unsupervised initial labeling and tracking. The moving object(s) as well as their overlapping characteristics were modeled using the concept of NRS. The uncertainty measures: neighborhood rough entropy, neighborhood probabilistic entropy and intuitionists entropy, defined over NRS, enabled quantifying the ambiguities.

The NRFT method (Chakraborty and Pal, 2018), described in this chapter, involves the modeling of moving object(s) and background as neighborhood rough set (NRS) over the natural neighborhood granules, described in Chapter 3 (see Section 3.3.1). The similarities in both color and spatial feature spaces are considered during this granulation, and the granules thus formed make the processing faster. An NRS filter was designed in this granulated space for initial labeling of the objects. Formation of two types of temporal neighborhood granules, namely, velocity and acceleration granules, are described here with the output of the filter to model the nature of the motion of the object(s) (Chakraborty and Pal, 2018).

As stated before, the NRFT method explores the merits of neighborhood rough set while modeling the video sequence as an NRS. The moving object estimations are done here with the concepts of lower and upper approximation of NRS. Whereas, in the NRBFG method, described in Chapter 3, the uncertainties present in the task of tracking were modeled using neighborhood rough rule-base and not with any lower or upper approximation.

Given an input video sequence, spatio-color neighborhood granules are formed over all the frames in NRFT. Its initial P number of frames are used for initial object labeling with the NRS filter developed in Chakraborty and Pal (2018). The color model and velocity profiles of the objects are the outputs of this filter from which the velocity and acceleration granules are generated. The object location in the next frame is estimated based on this information and the roughness in estimation is checked. If it is high, then the intuitionistic entropy is computed for the boundary granules and the ones which are found ambiguous are eliminated in order to track the object in the frame accurately. The tracking results, thus produced, are comparable to those of some of the state-of-the-art algorithms, while superior in case of total occlusion. These features have been demonstrated extensively in the task of tracking continuous moving object(s) in static background under different kinds of changes in shapes/sizes of moving object(s), changes in speed, multiple moving objects moving in different directions, occurrence of object-to-object overlapping, and object-to-background occlusion.

The chapter is organized as follows. Section 4.2 discusses some of the investigations on tracking overlapping occlusion of objects. The NRFT method along with its features is explained in details in Section 4.3. This includes the overall block diagram of tracking and the underlying concepts, and formulation of the theories of spatio-color granules, neighborhood rough filter, temporal granules and intuitionistic entropy. Section 4.4 describes the algorithms for tracking. In Section 4.5, all these characteristics and their effectiveness are experimentally demonstrated along with several comparisons. The overall conclusion of this chapter is drawn in Section 4.6.

4.2 Related Work on Handling Overlapping and Occlusion

Problems of multiple object tracking and handling their overlapping have recently become a challenging task in video tracking and analysis. The problem of multiple object tracking was primarily addressed by Raid (1979) where the current state is estimated from previous frames using Kalman filter. Later, particle filtering (also known as sequential Monte Carlo) was introduced, where a set of weighted particles sampled from a proposal distribution was maintained to represent the current and hidden states (Okuma *et al.*, 2004; Shabat *et al.*, 2015). This allows handling nonlinear multimodal distributions.

Kernel-based tracking (Comaniciu *et al.*, 2003) is another popular approach in visual tracking of nonrigid objects. Here the feature, namely, 'histogram-based target representation' is regularized by spatial masking with an isotropic kernel. Recently, a new kernelized correlation filter (KCF) (Henriques *et al.*, 2015) has been derived to handle the problems of multi-target tracking and occlusion. Unlike other kernel algorithms, it has the same complexity as its linear counterpart.

Sparse-based representation of object templates has become quite a popular approach to handle the ambiguities in videos. A sparse weight constraint is introduced in Hu *et al.* (2015) to dynamically select the relevant ones from the full set of templates. An algorithm for tracking multi-objects with occlusion based on the multi-feature joint sparse reconstruction is described over these templates. In Wang *et al.* (2015) an object is represented by the sparse coefficients of local patches based on an over-complete dictionary, and a classifier is learned to discriminate the target object from the background. In Bai *et al.* (2015) the sparse representation and online dictionary learning are unified by defining a sparsity consistency constraint that facilitates the generative and discriminative capabilities of the appearance model. An elastic-net constraint is enforced during the dictionary learning stage to capture the characteristics of the local appearances that are insensitive to partial occlusions.

There are several other approaches aimed to solve the aforesaid problems. The methodology developed in Park *et al.* (2015) is in the form of a binary integer programming problem where a polynomial time solution is provided that can obtain a good relaxation solution of the binary integer programming. This is then applied for multi-target tracking. In Pernici and Bimbo (2014) an object representation based on the weakly aligned multi-instance local features is defined which demonstrates that this representation improves on the inherent limit of local features invariance under occlusion. An energy function is developed in Milan *et al.* (2014) for tracking considering some physical constraints, such as target dynamics, mutual exclusion and track persistence. In addition, partial image evidence is handled with explicit occlusion reasoning, and different targets are disambiguated with an appearance model. The issue of re-identification of a person after partial or full occlusion was addressed by Hou *et al.* (2019) where Spatio-Temporal Completion network was developed to explicitly handle partial occlusion problems. Fernández-Sanjurjo *et al.* (2019) developed a solution to handle occlusion in traffic surveillance by combining deep learning-based detector and tracking through a combination of Discriminative Correlation Filter and a Kalman Filter.

The limitation of all these methods is that none of them can work without initial labeling of the object of interest. Besides, these methods may fail in detecting the moving object(s) when they are totally occluded by the background. In this context, the significance of the aforementioned NRFT method is evident, as it does not need any initial manual labeling, i.e., the labeling is unsupervised and is done automatically. The method has proved to be effective in handling even the cases like total occlusion or overlapping. Detailed characteristics of NRFT are explained in the following sections.

4.3 Unsupervised Methodology for Handling Overlapping and Occlusion

4.3.1 *Overview*

A brief overview of the NRFT method for unsupervised tracking is provided in Fig. 4.1. The method has two parts, namely: initial labeling and tracking. Initial P number of frames of a sequence are the input for initial labeling. Spatio-color granules are formed over these frames as shown in block A. Two types of frame-to-frame differences (viz, between the consecutive frames and from the Pth to all its previous frames as shown in block B) are computed in the granulated space. NRS-filter-based object estimation is then performed (block C), and the lower–upper sets of objects are formed (block M1). The temporal granules, namely, the velocity and acceleration granules are

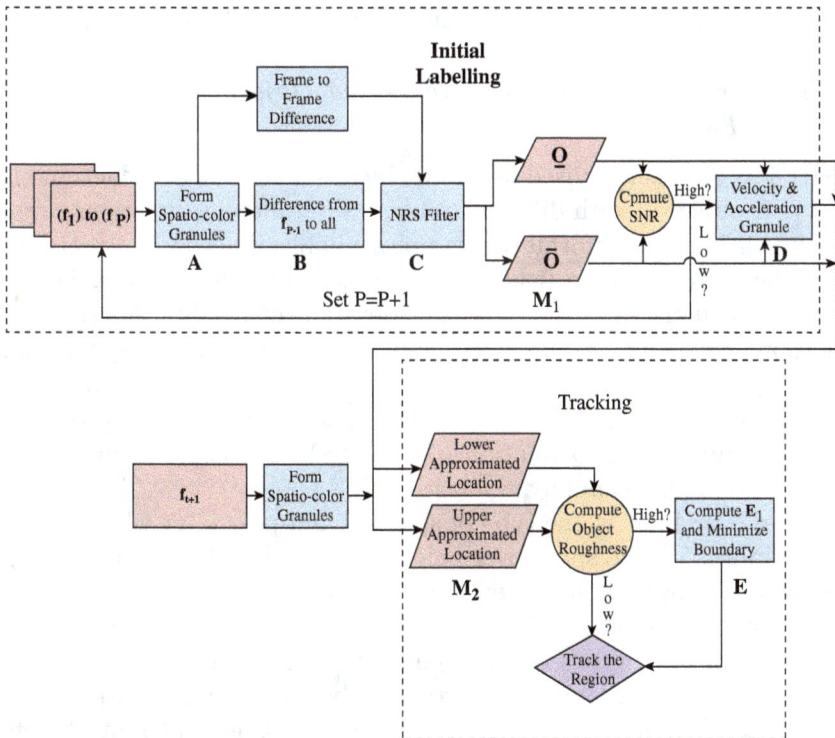

Fig. 4.1: Block diagram of the NRFT method.

generated using these estimated object sets (block D). This information is utilized during the task of tracking in the next frame (f_{t+1}) onwards. As shown in Fig. 4.1, the input frame to be tracked is first granulated and the locations are estimated on the frame (block M2) and the roughness in estimation is checked. If the roughness is high, then the intuitionistic entropy is computed for the boundary granules (block E) to ensure their presence in the object set, after which the object set is tracked.

These blocks are described in detail in the following sections.

4.3.2 *Formation of Spatio-color Neighborhood Granules*

As in the previous chapter, the spatial and color nearness are incorporated to form spatio-color granules in each input video frame (block A). The spatio-color neighborhood granules ($\aleph_{sp-clr}(x)$) are formed following the same steps, as described in Section 3.3.1 and Eq. (3.1).

4.3.3 *NRS Filter for Unsupervised Object Estimation*

NRFT is an unsupervised method which is able to detect continuous moving object(s) with different velocities in different directions, even if the background information is not available. That is, the decision-making is to be done from an incomplete knowledge base. Under this ambiguous situation, the consideration of neighborhood information of data points has been helpful in the process of estimation. A neighborhood rough set-based filter (neighborhood rough filter) was designed for this task where the concepts of lower approximation and upper approximation of object(s) were used to model the estimation approximately.

The NRS filter is one type of spatial band-pass filter which works in temporal feature space over neighborhood granular domain. Here, NRS-based lower and upper approximations of each object using the respective set of previous frames are performed. The convolution between the two types of changed information is executed to label which temporal segment will probably belong to which object in a frame. The different velocities and color models for different objects in a sequence are obtained from the output of this filter. Since the entire operation is carried out in the neighborhood granular domain, the filter is named as neighborhood rough filter. The principle of functioning of the filter is as follows.

The input to this filter consists of two types of temporal information corresponding to each frame. Let t denote the current frame and P be the total number of its previous frames. Then these two types of temporal information are:

(a) *Difference values between the consecutive frames* (δ_i) in granular level, i.e.,

$$\delta_i = \{\aleph_{sp-clr}(x) : |f_i(\aleph_{sp-clr}(x)) - f_{i-1}(\aleph_{sp-clr}(x))| > \mathcal{T}_1,$$
$$i = t, \ldots, t - P\}. \tag{4.1}$$

In Eq. (4.1), f_i and f_{i-1} represent two consecutive frames and $\mathcal{T}_1$ is the temporal threshold to detect the changed regions between frames.

(b) *Difference values from the current frame to all its previous frames* (δ_p) in granular level:

$$\delta_p = \{\aleph_{sp-clr}(x) : |f_t(\aleph_{sp-clr}(x)) - f_i(\aleph_{sp-clr}(x))| > \mathcal{T}_2,$$
$$i = t - 1, \ldots, t - P\}. \tag{4.2}$$

In Eq. (4.2), $\mathcal{T}_2$ is another temporal threshold to roughly detect the changed regions from current to all its previous frames.

These two pieces of information in the filter get convolved resulting in two types of approximated regions of objects in the current frame, corresponding to each previous frame. This is explained as follows.

Given a current frame and its P previous frames, the inputs to the filter are: (i) the union of the changed regions among current to all its previous frames ($\delta_P = \bigcup \delta_p : p = t - 1, \ldots, t - P$) and (ii) the change regions between the consecutive frames (δ_i). That is, a 'one to P points convolution' will take place in the filter which is expected to result in a P point matrix. Since two approximated decision spaces (viz, lower and upper) over the filter are defined to deal with the uncertainties, there will be two P point output matrices. The design of the filter is characterized by

$$\{\delta_i : i = 1, \ldots, P - 1\} * \delta_P = \{\underline{O}_c : c = 1, \ldots, P\}$$
$$\{\overline{O}_c : c = 1, \ldots, P\} \tag{4.3}$$

where the left-hand side and the right-hand side represent the input and output of the NRS filter. The two convolution parameters that were used during the approximation are as follows:

$$U_c = \delta_P \cup \{\delta_i : i = 1, \dots, P\} \quad \text{and} \tag{4.4}$$

$$I_c = \delta_P \cap \{\delta_i : i = 1, \dots, P\}. \tag{4.5}$$

$\underline{O}_c$ and $\overline{O}_c$ in Equation (4.3) represent the lower and upper approximated decision spaces, respectively, and are defined as follows:

$$\underline{O}_c = \{\aleph(x) \in U : \aleph(x) \in I_c \cap U_c\} \quad \text{and} \tag{4.6a}$$

$$\overline{O}_c = \{\aleph(x) \in U : \aleph(x) \in U_c \quad \& \quad \aleph(x) \cap I_c \neq \emptyset\}. \tag{4.6b}$$

The neighborhood granules ($\aleph(x)$), with which the approximations are performed, are the spatio-color granules (as described in Section 4.3.2) belonging to the regions δ_P and δ_i. An example of input and output of NRS filter is shown in Fig. 4.2 where $P = 3$. The convolution results in two 3×1 matrices representing two types of approximated regions of the objects, corresponding to each previous frame, as the output of NRS filter.

As seen, the filtering output over P frames roughly estimates the approximate locations and color models of objects. Based on this information, velocity profiles of objects are generated in order to

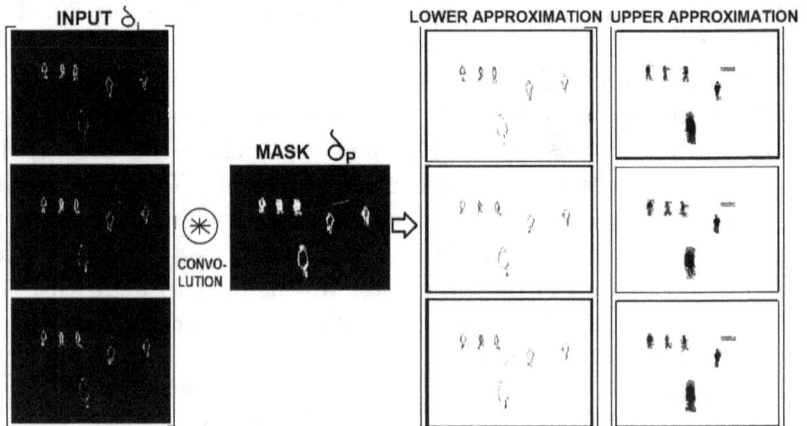

Fig. 4.2: An example of input and output of NRS filter.

estimate the locations of objects in the next frame. This is explained in Section 4.3.5. Before that, we describe the performance of NRS filter in Section 4.3.4.

4.3.4 *Performance Measure of NRS Filter*

The performance of NRS filter is measured in terms of the ratio of the cardinality of the obvious object sets (signal) to that of the boundary sets (probable noise). The cardinality of the object (boundary) sets is viewed as the amplitude of the signal (noise) sets therein.

The SNR of the NRS filter is defined as follows:

$$SNR = 20log_{10}\frac{|\underline{O}_c|}{|\overline{O}_c| - |\underline{O}_c|}. \tag{4.7}$$

The lower value of SNR reflects the poorer performance of the filter. The performance of the filter depends on its three parameters, viz, P, $\mathcal{T}_1$ and $\mathcal{T}_2$. These may be adjusted as follows:

If the value of $|\underline{O}_c|$ is very low, then it can be assumed that smaller than the exact region is extracted out from δ_i and hence the value of threshold $\mathcal{T}_1$ should be lowered down.

If the noise $|\overline{O}_c| - |\underline{O}_c|$ is very high, then it can be assumed that the extracted changed information from δ_P is more than that was required and the value of threshold $\mathcal{T}_2$ needs to be increased.

If neither the signal nor the noise deviates much from their expected values and SNR is low, then the number of previous frames P should be increased by two to have more information, and the entire process is repeated.

4.3.5 *Temporal Neighborhood Granules and Location Estimation*

The approximated locations and color models of the objects are roughly estimated after filtering over P consecutive frames (Sections 4.3.3 and 4.3.4). Using this information, the change in location of the object(s) in the next frame can be identified. Modeling of the nature of variation in size, speed and direction of the objects is therefore required to serve the purpose. The velocity neighborhood granules and the acceleration neighborhood granules are accordingly defined. These are discussed in the following sections.

4.3.5.1   *Velocity Neighborhood Granules* ($\aleph_V$)

These granules were formed by integrating the velocity information of objects, i.e., the changes of object locations from frame-to-
frame. It is known that velocity has two components: magnitude
and direction. The signed differences between the locations (primarily the locations of the four corners of object silhouette) in consecutive frames have been treated as the velocity in NRFT method.
Let $(x_1, y_1), \dots, (x_P, y_P)$ be the P positions of a corner of an object
$(\overline{O}_c : c = 1, \dots, P)$ appearing in P consecutive frames. The shifts
between the locations $\hat{V} \equiv (v_x, v_y)$ in consecutive frames are computed as follows:

$$v_{x_c} = x_c - x_{c-1} \tag{4.8a}$$

$$v_{y_c} = y_c - y_{c-1}, \quad c = 1, \dots, P. \tag{4.8b}$$

Here the signed difference is computed in Equation (4.8) to reflect
the direction.

Let the set of the shift values obtained from Eq. (4.8) be repre-
sented as $S_V = \hat{V}_c, \forall c = 1, \dots, P$. There are four such sets for the four
corners of each object. Velocity granules ($\aleph_v$) are constituted by the
shift values within the previous P number of frames which are nearer
to that of the previous frame, as follows:

$$\aleph_V = \hat{V}_i \in S_V : \triangle(\hat{V}_c, \hat{V}_t) < Th_V, \forall c = 1, \dots, P \tag{4.9}$$

There are four $\aleph_V$s for each object. The nearness threshold Th_V
was chosen based on the acceleration granules, discussed in the next
section.

These velocity granules ($\aleph_V$s) are able to separate the moving
object(s) (moving with consistency) and noise (random movement)
based on their cardinality. For example, the granules that are formed
over noise should be of cardinality one, as there is no continuous shift
for them.

4.3.5.2   *Acceleration Neighborhood Granule* ($\aleph_A$)

The moving object(s) in a video sequence do not usually move in a
uniform speed. Rather, the changes in shape, size, speed, direction,
etc. are obvious. The acceleration granules were defined to estimate

the pattern of the change in velocity in Chakraborty and Pal (2018). Let $(v_{x_1}, v_{y_1}), \ldots, (v_{x_P}, v_{y_P})$ be the P velocities of a corner of an object appearing in P consecutive frames. The changes in velocity of that corner, $\hat{A} \equiv (a_x, a_y)$, in consecutive frames are defined as follows:

$$a_{x_c} = v_{x_c} - v_{x_{c-1}} \tag{4.10a}$$

$$a_{y_c} = v_{y_c} - v_{y_{c-1}}, \quad c = 1, \ldots, P. \tag{4.10b}$$

The acceleration granules are formed over the set $S_A = \{\hat{A}_c, \forall c = 1, \ldots, P\}$ as

$$\aleph_A = \hat{A}_i \in S_A : \triangle(\hat{A}_c, \hat{A}_t) < Th_A, \forall c = 1, \ldots, P. \tag{4.11}$$

Here, Th_A is the nearness threshold, and there are four $\aleph_A$s for each object as in the case of $\aleph_V$.

The velocity and acceleration granules, thus formed for a frame, were used to estimate their probable values in the next frame. The search space in a frame is optimized with the help of these two types of granules once the object(s) are approximately known. Let the location of an object in the tth frame be L_t, then the predicted location of that object with $|\aleph_V| > P/2$ in the $(t+1)^{th}$ frame will be

$$\widetilde{L_{t+1}} = L_t \bigoplus \{\aleph_{V_r} : r \in \{RU, LU, LL, LC\}\}$$

$$\bigoplus \{\aleph_{A_r} : r \in \{RU, LU, LL, LC\}\} \in \overline{O} \tag{4.12}$$

where RU, RL, LL and LU represent the four corners of an object region and $\bigoplus$ stands for set summation. Note that, the region $\widetilde{L_{t+1}}$ is supposed to contain the set of spatio-color neighborhood granules in $\overline{O}$. Here $\overline{O} = \{\overline{O}_c : c \in P\}$.

An example of location estimation of the objects (*obs*) over the lower approximated output of Fig. 4.2 is shown in Fig. 4.3 where the estimated locations are shown in gray color. As seen in Fig. 4.3, there is no estimation performed for the object 'ob7' as there is no continuous shift in this region. It is therefore detected as noise.

4.3.6 *Roughness in Object Approximation*

Let there be $\mathcal{M}$ number of moving objects in a sequence. Then, $\mathcal{M}$ number of estimated locations $\widetilde{L_{t+1}}$ will be there in the tth frame.

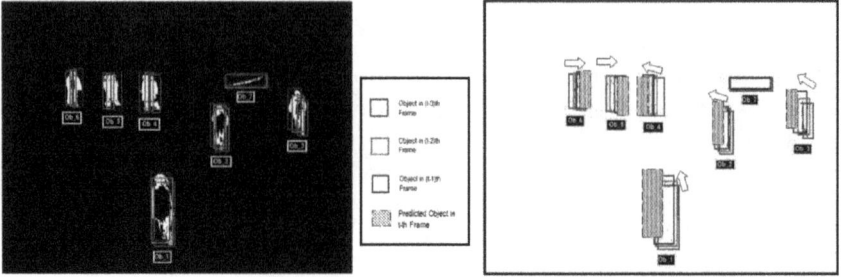

Fig. 4.3: Example location estimation over the output from NRS filter.

The roughness of the estimated locations is measured based on how many similar color granules ($\aleph_{clr}$) are there in the approximated regions in the current frame as compared to that of the object models. Let the approximated model of the mth moving object be $\overline{\mathcal{O}}_m = \{\cup \aleph_{clr} : \aleph_{clr} \in (\overline{O_{c_m}} \cap \overline{L_{t_m}} \forall c = 1, \ldots, P)$. Then the roughness of the object model approximation ($\tilde{R\mathcal{O}}_m$) is defined as follows:

$$\tilde{R\mathcal{O}}_m = 1 - \frac{|\tilde{L_m}(\aleph_{clr}) \cap \overline{\mathcal{O}}_m(\aleph_{clr})|}{|\overline{\mathcal{O}}_m(\aleph_{clr})|}. \tag{4.13}$$

In Eq. (4.13), $|\tilde{L_m}(\aleph_{clr}) \cap \mathcal{O}_m(\aleph_{clr})|$ represents the total number of similar color granules between the estimated object models and the estimated regions. If the value of $\tilde{R\mathcal{O}}_m$ is high (as determined by a predefined threshold Th_O, say) for the mth object, it means that there are less number of similar granules. In other words, it can be hypothecated that there occurs some occlusion over that object either with the background or with another object. The concept of intuitionistic entropy measure was, accordingly, introduced by Chakraborty and Pal (2018) to deal with such issues by estimating whether a granule is intuitionally present in the object set in terms of its overlapping nature and its probability of being in the predicted location. The details of it are described in the following section.

4.3.7 *Intuitionistic Entropy*

This entropy was defined by incorporating the merits of both rough entropy and probabilistic entropy. The granules over which this measure was performed are those which are present in the object model ($\mathcal{O}_m$), but not completely in the predicted region ($\mathcal{O}_m(\aleph_{clr}) - \tilde{L}_m(\aleph_{clr}) \cap \mathcal{O}_m(\aleph_{clr})$). The presence of a spatio-color granule in the predicted object region was determined even if it was visually absent. This entropy was constituted by two types of uncertainty measures: neighborhood rough entropy and neighborhood probabilistic entropy. The first one quantifies the importance of a boundary granule to the object model, whereas the second one evaluates its probability of being in the predicted region. In this way the combination of these two measures intuitionally estimates the object in the next frame. The neighborhood rough entropy, formulated in Chakraborty and Pal (2018), incorporates the characteristics of overlapping, whereas the neighborhood probabilistic entropy ensures the estimation.

4.3.7.1 *Neighborhood Rough Entropy*

The importance of a boundary granule $\aleph_{BO}$ to the respective object model ($\mathcal{O}_m$) is measured with this entropy. The cardinality measure of a set was redefined in this process as NRS was dealing with overlapping granules. This implied that same points may belong to more than one granule. Therefore, instead of counting each granule as an individual entity to measure the cardinality of an approximated set and considering its full membership either to lower or upper approximation, like Pawlak's rough sets PaRS (Pawlak, 1992), its partial membership to those approximations is computed here while measuring the roughness. The overlapping cardinality (represented as $\langle . \rangle$) of a neighborhood rough set (N) is defined as

$$\langle N \rangle = \sum_{g \in G} \partial_g \qquad (4.14)$$

where G is the total number of granules in the set N. ∂_g denotes the part of the granule not already counted in other approximations, as defined in Eq. (4.17).

The neighborhood rough entropy (NRE) that is defined here concerns only with one (object) class, as the background class is not estimated. It is expressed as

$$NRE = -e * RO * \ln(RO) \qquad (4.15)$$

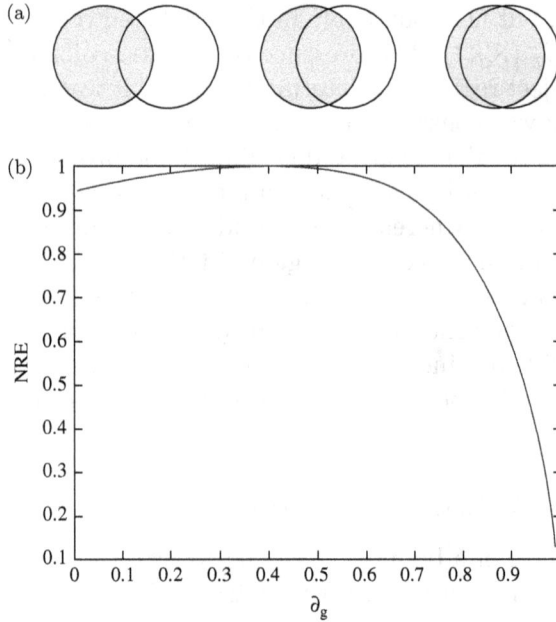

Fig. 4.4:   Example of two overlapping granules (a) granules with different degrees of overlapping and (b) variation in NRE with ∂_g.

where RO denotes the roughness of an object class overlapping with a boundary granule $\aleph_{BO}$. It is defined as

$$RO = 1 - \frac{\langle \mathcal{O}_m - \aleph_{BO} \rangle}{\langle \mathcal{O}_m \rangle}. \tag{4.16}$$

NRE reflects the amount of difficulty in deciding whether a boundary granule $\aleph_{BO}$ belongs to the object region or not (noise). An example is shown in Fig. 4.4 to demonstrate the effect of ∂_g on NRE where a simple case with two granules overlapping with each other is considered. Let us assume that the gray granule in Fig. 4.4(a) is in the lower approximation and the white one is the boundary granule. Fig. 4.4(a) shows that the NRE value for a boundary granule $\aleph_{BO}$ is high if there is minimum overlapping ($0 \leq \partial_g < 0.5$). It attains the maximum value when $\partial_g = 0.5$ as the decision-making over that boundary granule becomes the toughest in this scenario. The NRE decreases gradually after $\partial_g > 0.5$ as the degree of overlapping increases, and it attains its minimum value ($= 0$) once $\partial_g = 1$, that is the lower and upper approximations are the same and no more uncertainty is there. Therefore, the granules with $\partial_g < 0.5$ may be considered as noise. Note that had the conventional definition of cardinality as in Pawlak's rough set been used, the entropy value would have been the same for any value of ∂_g.

NRE determines the significance of $\aleph_{BO}$ to the object model. Now its probability of being present in the estimated region $\tilde{L}_{t+1}$ is determined by neighborhood probabilistic entropy, as discussed in the following section.

$$\partial_g = \frac{\text{Points not in overlapping regions with already counted granules}}{\text{Total nos. of points in the granule}}.$$

$$(4.17)$$

4.3.7.2 *Neighborhood Probabilistic Entropy*

This ambiguity measure was formulated based on the probability of occurrence of a granule, and not on the cardinalities of the approximated sets. Two types of probabilities: (i) the probability of occurrence of $\aleph_{BO}$ over the sequence and (ii) its probability of being in the $\tilde{L}_{t+1}$ are considered. The computation of the second one is dependent on the first one.

Let $\aleph_{BO}$ be present $\mathcal{P}$ times in the object models obtained over the previous P frames (i.e., $\mathcal{P} \leq P$). Then its probability of occurrence (NPO) over the sequence will be

$$NPO = \frac{\mathcal{P}}{P}.$$

$$(4.18)$$

Let the location of $\aleph_{BO}$ in $(t)^{th}$ frame be l_t and its predicted location in the current frame be $\tilde{l}_{t+1} = l_t \oplus v_m \oplus a_m$ according to Equation (4.12), where v_m and a_m are the mean of $\aleph_V$ and $\aleph_A$, respectively. The probability of being the granule in the predicted region (NPR) is computed based on the belongingness of $\tilde{l}_{t+1}$ in $\tilde{L}_{t+1}$ over NPO (Eq. (4.18)). Therefore it is a conditional probability which is described as

$$NPR = NPO \times \frac{|\tilde{l}_{t+1} \cap \tilde{L}_{t+1}|}{|\tilde{l}_{t+1}|}.$$

$$(4.19)$$

The neighborhood probabilistic entropy (NPE) is defined as

$$NPE = -e \times NPR \times \ln(NPR).$$

$$(4.20)$$

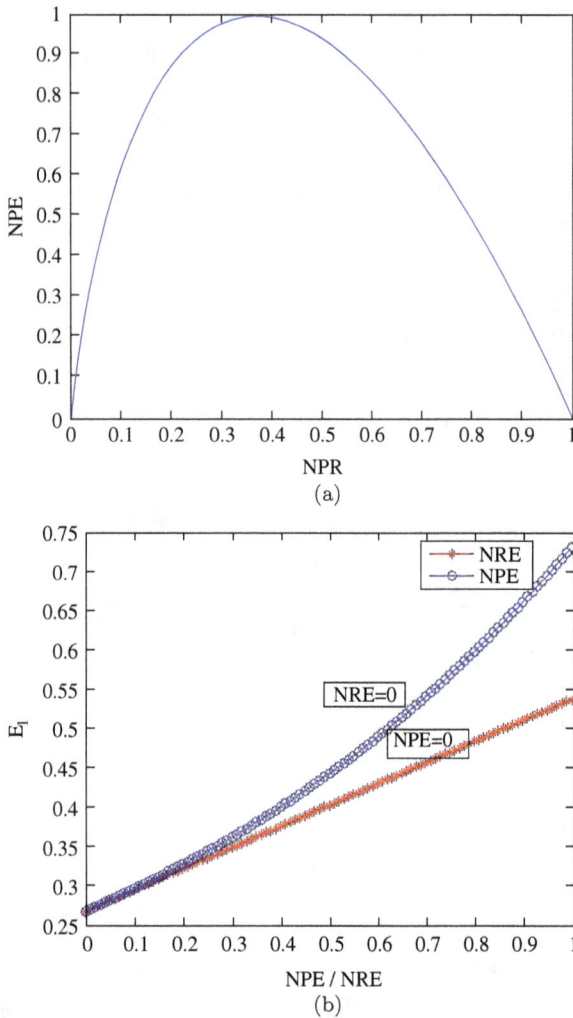

Fig. 4.5: Plot of (a) *NPR* vs. NPE and (b) variation of E_I with NPE and NRE.

The characteristics of *NPE* vs. *NPR* is shown in Fig. 4.5. It can be noticed that *NPE* increases initially with *NPR* and attains its maximum value at $\frac{1}{e}$, then it starts to decrease. That is, the granules with $NRE < \frac{1}{e}$ may be treated as noise and not considered as probable object granules.

The intuitionistic entropy (E_I) considers the effects of both *NRE* and *NPE*. E_I is measured twice for each boundary granule $\aleph_{BO}$

(with respect to object set and that of its complement) to ensure its presence in the object set. Let NRE_1 be the value of NRE, and NPE_1 be the value of NPE after inserting $\aleph_{BO}$ in the Q. The object intuitionistic entropy (E_{I_O}) of the granule was defined as

$$E_{I_O} = \frac{NRE_1 + e^{NPE_1}}{e + 1} \tag{4.21a}$$

Similarly, the background intuitionistic entropy E_{I_B} was defined with the complement values of NRE_1 and NPE_1, denoted as NRE_1^c and NPE_1^c, respectively. That is

$$E_{I_B} = \frac{NRE^c + e^{NPE^c}}{e + 1} \tag{4.22a}$$

If $E_{I_B} > E_{I_O}$, it means, the boundary granule $\aleph_{BO}$ reduces the overall ambiguity when it is in the object set. Then $\aleph_{BO}$ is labeled as an object granule, otherwise the background (noise).

The nature of variation of E_I with NPE and NRE is shown separately in Fig. 4.5(b). The red 'o−' line shows the variation of E_I with NPE, which is exponential in nature, and the blue '*−' line shows its variation with NRE, which is linear. That is, the impact of the change in NPE on E_I is higher as its probable deviation is taken into account during the formulation. This is not the case for NRE where the impact is linear.

4.4 Algorithm for Tracking

The overall algorithm for NRFT method is described here in brief for a single moving object. This algorithm primarily consists of two parts: training and testing. These are described in Algorithms 2 and 3, respectively.

The same steps of these algorithms will be repeated M times for M number of moving objects in a sequence. The complexity of all these algorithms is given in Chakraborty and Pal (2018).

4.5 Experimental Results

The effectiveness of the NRFT algorithm is shown in this section with some experimental results (Chakraborty and Pal, 2018). The key features/advantages of this algorithm lie in:

Algorithm 2 Unsupervised Object(s) Model Estimation with NRS-filter

---

INPUT: Initial P frames, Spatio-color granular threshold Th, Temporal thresholds: $\mathcal{T}_1$, $\mathcal{T}_2$

OUTPUT: Approximated object models $\underline{O}_c$ and $\overline{O}_c$: $c = 1, ..., P$

INITIALIZE: $\underline{O}_c = \overline{O}_c \Leftarrow \emptyset$

1: Compute temporal values δ_i and δ_p.

2: Form spatio-color neighborhood granules (Section 4.3.2) over P frames.

3: Perform rough object–background separation with NRS filter (see Section 4.3.3) to extract out $\overline{O}_c$ and $\underline{O}_c : c = 1, ..., P - 1..$

4: Measure the performance (SNR) of it as described in Section 4.3.4.

if SNR is low **then**

    while $\overline{O}_c - \underline{O}_c < \underline{O}_c$ **do**

        set $P = P + 2$; go to step 1.

    end while

end if

5: Approximate the objects' models $\overline{O}$.

6: Form velocity and acceleration granules ($\aleph_V$, $\aleph_A$) (see Section 4.3.5).

(i) Tracking multiple objects, without any prior knowledge, even with variations in shape/size or speed, and

(ii) Tracking object(s) even if it is fully/partially occluded with background or gets overlapped with other moving object(s).

Video sequences with different characteristics, e.g., indoor/outdoor surveillance (Davis and Sharma, 2007; PETS-2001, 2001; PETS-2009, 2009; AVSS-2007, 2007; Song and Xiao, 2013), single/multiple moving object(s) of Ordinary Videos for evaluating visual trackers robustness (2014); Possegger *et al.* (2013); Wu *et al.* (2015); PETS-2015 (2015), body part(s) movements (ChaLearn, 2011) were considered during the experiment. The algorithm was executed almost over 2500 frames in total. The sequences over which the results are shown here, as examples, are described below. 2*b*-sequence (Davis and Sharma, 2007) has six people moving in different directions. Sequence M_10 (ChaLearn, 2011) has one lady who is moving her hands with overlapping between them. In the sequence $14 - LD$ of Ordinary Videos for evaluating visual trackers robustness (2014), one person is moving with total

---

Algorithm 3 Object Tracking

INPUT: Next frame f_{t+1}, $\overline{\mathcal{O}}$, $\aleph_V$, $\aleph_A$, Th_o

OUTPUT: Object in the current frame O

INITIALIZE: $O \Leftarrow \emptyset$

1: Approximate locations $\overline{\tilde{L}_{t+1}}$ in the next frame.

2: Compute the roughness in object location approximation $\tilde{R\mathcal{O}}$.

if $\tilde{R\mathcal{O}} > Th_o$ **then**

  Compute intuitionistic entropy and estimate $\overline{\tilde{L}_{t+1}}$.

else

  Set $\overline{O}_{t+1} = \overline{\tilde{L}_{t+1}}$.

end if

3: Track $\overline{\tilde{L}_{t+1}}$.

4: Update $\aleph_V$ and $\aleph_A$,

5: Update $\overline{\mathcal{O}}$.

6: Set $t + 1 = t$.

7: Repeat step 1 to step 4.

---

occlusion twice. There are five people moving in a group, get occluded to each other and the background in the sequence $P - 15$ (PETS-2015, 2015). There are three moving people who get occluded and overlapped with one another in the sequence New_Occ4 (Song and Xiao, 2013). In the sequence $S2 - L1$ (PETS-2009, 2009), multiple people are moving in different directions, get overlapped, and new persons appear. Sequence $P - 01$ (PETS-2001, 2001) initially contains one moving person and then one moving car appears at different velocities and with different variations in shape and size over the sequence. In $A - 07$ sequence (AVSS-2007, 2007), there is initially one person moving, gets totally occluded with background and reappears, then another person appears from a different direction and moves. Two people enter one by one and get partially overlapped with change in their directions in the sequence $much - 132$ (Possegger *et al.*, 2013). The results obtained with these datasets are shown in the following sections.

4.5.1 *Selection of Parameters and Initial Assumptions*

The underlying assumptions during the initial labeling with NRFT method were: (i) no moving object should get occluded or overlapped and (ii) the moving object and its respective background should be well separable in RGB feature space. Selection of parameters is a crucial issue in every tracking algorithm. In NRFT method, the parameters were made adaptive, as much as possible, so that these can take the approximate values automatically depending on the nature of the movement and size of the objects in a sequence. The method of selecting the parameter P depending on the speed of the object(s) from frame-to-frame for the aforesaid sequences is explained in Section 4.5.2. The initial values of T_1 and T_2 were set as 15 during the experimentation in Chakraborty and Pal (2018), and they get updated according to the requirement of the respective video sequence. The value of Th was set as 30, as this value was experimentally found to provide a balanced trade off between the accuracy and number of segments.

4.5.2   *Effectiveness of Unsupervised Labeling and Estimation*

The results for initial unsupervised labeling of continuous moving object(s) using NRS filter are shown in this section. The visual results in Fig. 4.6 show the first labeled frames with respect to the appearance of the object(s) in the respective sequences. For example, no new object(s) appear in the sequences: $2 - b$, M_10, $14 - LD$, $P - 15$ and New_Occ4 (see Figs. 4.6(a), (b), (c), (d), (e)). But new moving object(s) appear in the sequences $S2 - L1$, $P - 01$, $A - 07$, $much - 132$ (see Figs. 4.6(f), (g), (h), (i)). The initially present object(s) in the respective sequences are shown in Figs. 4.6(f)-(i), (g)-(i), (h)-(i), and (i)-(i); and the object(s) which appeared late are shown in Figs. 4.6(f)-(ii), (g)-(ii), (h)-(ii), and (i)-(ii). The initially present object(s) (I) in each sequence are marked by red trackers whereas the newly appeared ones (N) are marked by violet trackers in Fig. 4.6.

The required number of frames (P) to label an object varies depending on the speed of the object even in the same sequence. This is shown in Table 4.1 where P-values required for labeling the initially present object(s) as well as the newly appeared object(s) for the aforesaid sequences are given. The accuracy of initial labeling is measured in terms of the root mean square distance between the four corners of bounding boxes of the annotated ground truth and that of

Fig. 4.6: Initially tagged objects of (a) $2-b$ sequence, (b) M_10 sequence, (c) $14-LD$ sequence, (d) $P-15$ sequence, (e) New_Occ4 sequence, (f) $S2-L1$ sequence, (g) $P-01$ sequence, (h) $A-07$ sequence and (i) $much-132$ sequence: (i) initially present object(s) and (ii) newly appeared object(s).

Table 4.1: Effectiveness of NRS filter-based unsupervised labeling.

Sequences	$2-$ b	$M\_$ 10	$14-$ LD	$P-$ 15	$New\_$ $Occ4$	$S2-$ $L1$-I	$S2-$ $L1$-N	$P-$ 01-I	$P-$ 01-N	$A-$ 07-I	$A-$ 07-N	$much-$ 132-I	$much-$ 132-N
P	10	7	8	11	5	6	7	17	9	7	8	7	6
$RMSE_L$	11.4	9.3	5.4	12.2	7.5	12.7	15.1	6.3	12.2	5.2	14.8	11.1	9.2
$RMSE_P$	10.7	7.11	4.3	11.4	6.8	14.2	13.4	5.9	11.4	6.4	13.9	9.2	15.1

the labeled object $(\overline{O}_c)$ $(RMSE_L)$. The other metric that is shown in the table is accuracy in prediction in the next frame $(RMSE_P)$ which is measured the same way as that of $RMSE_L$, where the distance between the predicted region in the next frame of the labeled one $(\widetilde{L}_t$ in Section 4.3.5) and the corresponding ground truth is computed.

It can be noticed from Table 4.1 that the parameter P was selected depending on the speed of the object and it varies for each object accordingly. Different types of objects in different scenarios are identified with $RMSE_L$ attaining values between 5.4–14.8 pixels for the aforesaid sequences. This can be interpreted as quite efficient, as evident from the results of tracking described in the next section. A similar conclusion can be drawn for the prediction task by observing the $RMSE_P$-values. These results demonstrate the effectiveness of the NRS filter, and the velocity and acceleration granules for the tasks of unsupervised object labeling and location estimation, respectively.

4.5.3 *Results of Tracking*

The effectiveness of NRFT algorithm in dealing with the tracking cases involving (i) variation in shape/size ($P-01$, $A-07$), (ii) appearance of new object(s) (as discussed in Section 4.5.2), (iii) non-uniform movement of object(s) ($S2 - L1$, M_10, $much - 132$) is shown here. The results for dealing with occlusion using the intuitionistic entropy are also shown. The datasets used here characterize different types of occlusions: (i) object-to-background ($14-LD$, $A-07$, $S2_L1$, $P-15$), (ii) object-to-object ($S2 - L1$, M_10, $much - 132$, New_Occ4), (iii) partial ($S2-L1$, M_10, $much-132$) and (iv) total ($14-LD$, $A-07$). The comparative studies were carried out with seven recent partially supervised methods (where the object(s) are manually labeled in the initial frame) which have proved to be effective and robust to multi-object tracking and occlusion handling demonstrating the merits of NRFT. These comparing methods along with their parameter values are described in what follows.

MTJSR (Hu *et al.*, 2015) where the templates from different image features are fused to form a robust object model and the tracking is done with particle filter. Here the Gabor filter was set to 2 and $\frac{\pi}{2}$ according to the article.

MCIO (Park *et al.*, 2015) where many to one (M to 1) optimization approach is performed for tracking multiple interacting objects. The parameters were set as: $M = 2$, $\sigma = 0.3$ and $g = 4$ following the article.

OM (Milan *et al.*, 2014) where an energy-based model of multi-target tracking is defined over all the target locations and all video frames in a given time window. Here the values of the parameters are: $\lambda = 0.1$, $\delta = \epsilon = 0.6$ and $\beta = 0.03$ as given in the paper.

ALIEN (Pernici and Bimbo, 2014) where the objects are represented by their invariant local features taken in different conditions (oversampling). Following the article the parameters were set as: $\lambda_T = \lambda_C = 2.5$, $l = 10$, $n_u = 8$ and $\sigma = 1$.

JORS (Wang *et al.*, 2015) where object tracking is formulated as an optimization problem by considering both the reconstruction and classification errors in the objective function. Here $N = 15$, $\beta = 0.2$, $T_1 = 0.05$, $T_2 = 0.5$ and $T_3 = 0.7$ as instructed in the article.

DeepTrack (Li *et al.*, 2016) where tracking is performed using a single convolutional neural network (CNN) which is developed for learning effective feature representations of the target object in a purely online manner. Here the input is locally normalized into 32×32 image patches in the first layer and 7×7 image patches in the second layer as that was there in the paper.

CNT (Zhang *et al.*, 2016) where simple two-layer convolutional networks are used for object representations and a 'deep network' structure is introduced in visual tracking. Here the size of the wrapped images and field size are set to 32×32 and 7×7, respectively, following the article.

The visual comparative results of NRFT with these seven other robust methods are shown in Fig. 4.7. It can be noticed from the figure that all of the algorithms perform well in multiple object tracking as well as for object-to-object overlapping/interaction handling. However, the NRFT algorithm works better when total object-to-background occlusion takes place as the intuitionistic entropy incorporates both of the overlapping and probabilistic properties of the moving object(s). It can be seen in Figs. 4.7(c-(iv)), (e-(iii)), (h-(ii), h-(iii)) that the unseen object(s) (totally occluded by background/other object) are almost correctly identified by the NRFT method, whereas there are localization errors for the other algorithms. The quantitative comparative studies in terms of average *RMSE* (computed similarly as that of *RMSE_L*) and percentage of correctly classified frames in each sequence are shown in Tables 4.2 and 4.3, respectively. The performances of all these eight methods throughout the aforesaid nine sequences are shown in Fig. 4.8. It is seen in Fig. 4.8 that all the eight methods perform well when there is no occlusion or overlapping, whereas the performances of the methods 'MTJSR','MCIO','OM','ALIEN','JORS' deviate and attain higher values when some overlapping takes place. One may note the

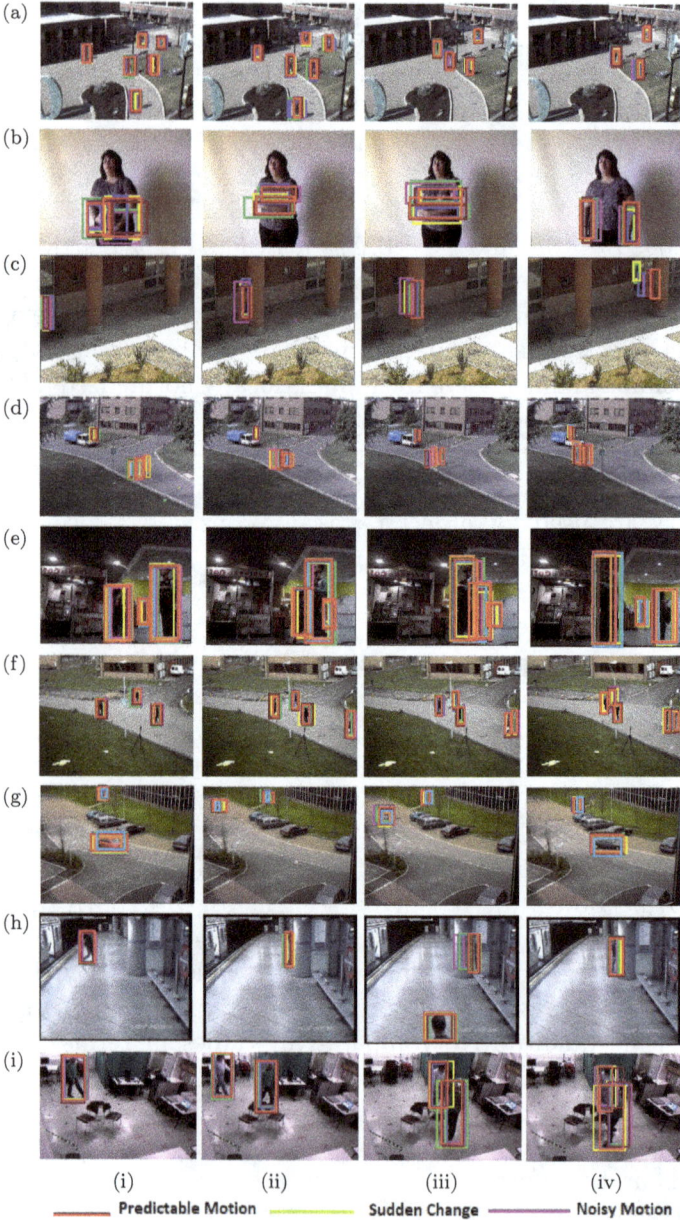

Fig. 4.7: Results of tracking for (a) $2 - b$ sequence, (b) M_10 sequence, (c) $14 - LD$ sequence, (d) $P - 15$ sequence, (e) New_Occ4 sequence, (f) $S2 - L1$ sequence, (g) $P - 01$ sequence, (h) $A - 07$ sequence and (i) $much - 132$ sequence.

Table 4.2:   Average RMSE.

Sequences	$2 - b$	M_10	$14 - LD$	$P - 15$	New_Occ4	$S2 - L1$	$P - 01$	$A - 07$	$much - 132$
$MTJSR$	3.2	2.4	11.2	2.6	5.8	4.1	1.8	8.2	2.6
$MCIO$	2.3	3.7	10.7	2.3	8.2	3.9	1.2	7.8	2.4
OM	0.8	3.25	8.3	1.8	8.9	3.65	2.4	7.9	3.01
$ALIEN$	0.56	3.45	5.2	1.7	6.1	3.76	2.2	8.1	3.2
$JORS$	3.26	4.15	12.3	4.1	9.7	3.33	1.72	12.1	2.22
$DeepTrack$	1.23	2.41	8.3	0.8	5.1	2.88	1.47	3.9	1.68
CNT	0.97	2.58	8.1	1.6	6.21	3.08	1.13	7.8	1.91
$NRFT$	3.11	3.35	3.98	0.83	2.67	3.13	1.22	3.1	2.13

Table 4.3:   % of correctly classified frames.

Sequences	$2 - b$	M_10	$14 - LD$	$P - 15$	New_Occ4	$S2 - L1$	$P - 01$	$A - 07$	$much - 132$
$MTJSR$	98	97	80	82	90	93	96	86	99
$MCIO$	99	96	82	85	95	98	99	85	97
OM	100	98	83	84	93	97	97	87	99
$ALIEN$	100	97	81	84	96	98	99	90	98
$JORS$	94	95	79	83	92	93	97	82	98
$DeepTrack$	92	96	88	90	96	95	97	91	99
CNT	93	95	92	93	95	91	94	96	98
$NRFT$	98	97	98	96	97	98	98	98	97

values of RMSE for the sequences M_10, $P-15$, $New-Occ4$, $S2-L1$ and $much - 132$ in Fig 4.8 (b), (d), (e), (f) and (i) deviated for the said methods over the frames where overlapping/occlusion takes place. Values of RMSE increase for the other two methods, namely, 'DeepTrack' and 'CNT', when total occlusion occurs. Total object-to-background occlusion takes place in the sequences $14 - LD$ and $A - 07$ (Fig. 4.8(c) and (h)). The comparing algorithms fail to estimate the totally occluded objects resulting in a higher RMSE value, however, they can again track those once they reappear (become visible). RMSE for the NRFT method remains almost the same throughout the sequences (though not the best for all the cases), when no new object appears. That is, the NRFT method is well capable of handling total occlusion or overlapping. These results conform to the visual results. The average $RMSE$ (see Table 4.2) values also reflect similar conclusions for all the sequences with multiple moving objects or with very low object-to-object interactions. This proves

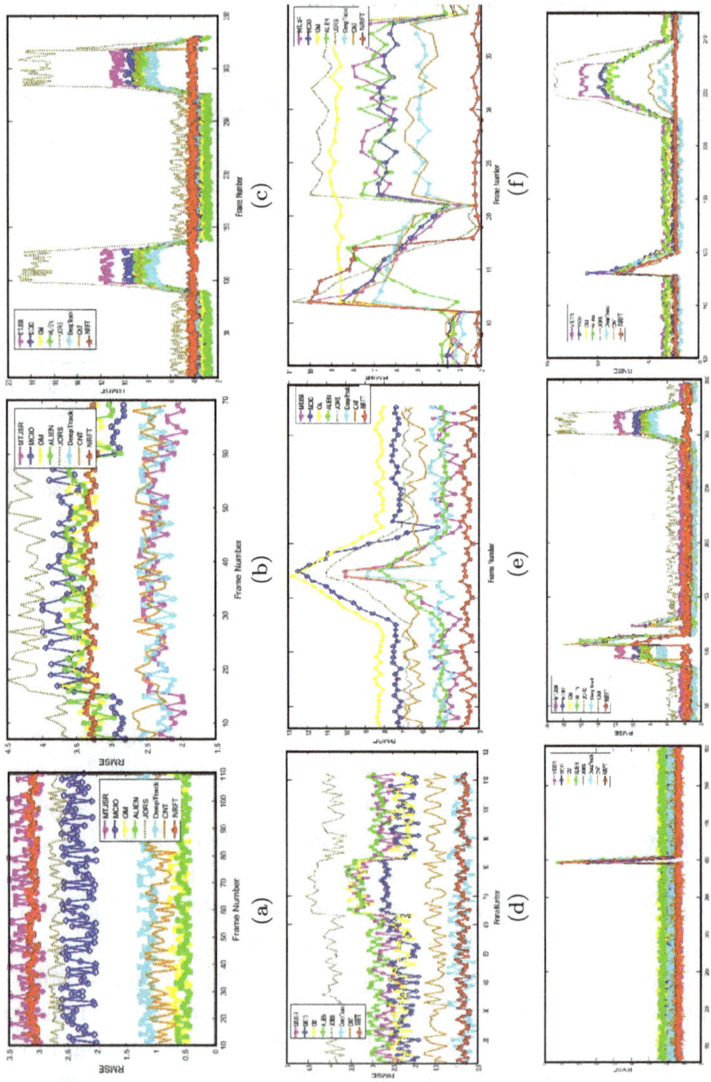

Fig. 4.8: RMSE value throughout the sequences: (a) $2 - b$ sequence, (b) M_10 sequence, (c) $14 - LD$ sequence, (d) $P - 15$ sequence, (e) *New_Occ4* sequence, (f) $S2 - L1$ sequence, (g) $P - 01$ sequence, (h) $A - 07$ sequence and (i) *much* $- 132$ sequence.

the effectiveness of all the comparisons as well as NRFT algorithms in handling such cases. The percentage of the correctly classified frames (see Table 4.3) obtained by these algorithms for the aforesaid sequences is very high reflecting their accuracy in tracking throughout the sequences. It can also be observed that OM works the best for multi-target tracking, whereas DeepTrack for interacting objects. The NRFT method can estimate the locations of the totally occluded object throughout with the accuracy ($RMSE$) value around 3.

4.6   Conclusions and Discussions

The task of handling occlusion in video tracking under unsupervised framework has been addressed in this chapter. The significance of rough sets and granular computing in handling uncertainties and predicting the locations of objects in this task is investigated. The recently developed neighborhood rough filter-based tracking (NRFT) method is discussed in detail. The NRFT provides a methodology for unsupervised tracking of multiple moving object(s) considering the cases like occlusions/overlapping by exploring the features of neighborhood rough sets. The NRS filter that was designed here in this method is able to label the continuous moving object(s) accurately without any manual interactions in terms of two (lower and upper) approximated decision spaces. The filter functions are in the spatio-color granular domain of video frames. The intuitionistic entropy enables one to handle efficiently the cases like partial/total occlusions/overlapping. This unsupervised NRFT method provides equally good results in the task of tracking when compared with the other recent partially supervised methods, and is proved to be superior in the frames where total object-to-background occlusion takes place.

The method can be further extended for real-time applications by exploring more merits of rough sets and granular computing. The concept of NRS filter and intuitionistic entropy can be applied to the other areas of signal processing where the unsupervised prediction and ambiguity handling are of concern.

We have so far discussed the impact of rough set theoretic granular computing in different variations of the primary component of video analysis, i.e., tracking of moving objects, under different ambiguous conditions. Now we will focus on some other challenges in video analysis. The problem of quantifying the trustability of different tracking methods under different conditions is addressed in the next chapter. The last two chapters describe the effectiveness of rough sets and granular computing in the tasks of object recognition from video scenes and event recognition from videos.

Bibliography

Amsterdam Library of Ordinary Videos for evaluating visual trackers robustness (2014). *ALOV300++ Dataset*.

AVSS-2007 (2007). *Fourth IEEE Int. Conf. Adv. Video & Signal Based Surveillance*.

Bai, T., Li, Y.-F. and Zhou, X. (2015). Learning local appearances with sparse representation for robust and fast visual tracking, *IEEE Trans. Cyberns.* **45**(4): 663–675.

Chakraborty, D. B. and Pal, S. K. (2018). Neighborhood rough filter and intuitionistic entropy in unsupervised tracking, *IEEE Trans. Fuzzy Systems* **26**(4): 2188–2200.

ChaLearn (2011). *ChaLearn Gesture Dataset (CGD 2011)*, California.

Comaniciu, D., Ramesh, V. and Meer, P. (2003). Kernel-based object tracking, *IEEE Trans. PAMI* **25**(5): 564–575.

Davis, J. and Sharma, V. (2007). Background-subtraction using contour-based fusion of thermal and visible imagery, *Comput. Vis. Image Underst.* **106**: 162–182.

Fernández-Sanjurjo, M., Mucientes, M. and Brea, V. M. (2019). Real-time traffic monitoring with occlusion handling, *Pattern Recognition and Image Analysis*, Springer International Publishing.

Henriques, J. F., Caseiro, R., Martins, P. and Batista, J. (2015). High-speed tracking with kernelized correlation filters, *IEEE Trans. PAMI* **37**(3): 583–596.

Hou, R., Ma, B., Chang, H., Gu, X., Shan, S. and Chen, X. (2019). Vrstc: Occlusion-free video person re-identification, *IEEE Conference on Computer Vision and Pattern Recognition (CVPR)*.

Hu, W., Li, W., Zhang, X. and Maybank, S. J. (2015). Single and multiple object tracking using a multi-feature joint sparse representation, *IEEE Trans. PAMI* **37**(4): 816–833.

Li, X., Shen, C., Dick, A. R., Zhang, Z. M. and Zhuang, Y. (2016). Online metric-weighted linear representations for robust visual tracking, *IEEE Trans. Pattern Anal. Mach. Intell.* **38**(5): 931–950.

Milan, A., Roth, S. and Schindler, K. (2014). Continuous energy minimization for multitarget tracking, *IEEE Trans. PAMI* **36**(1): 58–72.

Okuma, K., Taleghani, A., Freitas, O., Little, J. and Lowe, D. (2004). A boosted particle filter: Multitarget detection and tracking, *Proc. European Conf. Computer Vision*, Vol. 1, pp. 28–39.

Park, C., Woehl, T. J., Evans, J. E. and Browning, N. D. (2015). Minimum cost multi-way data association for optimizing multitarget tracking of interacting objects, *IEEE Trans. PAMI* **37**(3): 611–624.

Pawlak, Z. (1992). *Rough Sets: Theoretical Aspects of Reasoning about Data*, Kluwer Academic Publishers, Norwell, MA.

Pernici, F. and Bimbo, A. (2014). Object tracking by oversampling local features, *IEEE Trans. PAMI* **36**(12): 2538–2551.

PETS-2001 (2001). *IEEE Int. WS Perfor. Evaluation of Tracking and Surveillance*.

PETS-2009 (2009). *IEEE Int. WS Perfor. Evaluation of Tracking and Surveillance and EC Funded CAVIAR project*.

PETS-2015 (2015). *IEEE Int. WS Perfor. Evaluation of Tracking and Surveillance*.

Possegger, H., Sternig, S., Mauthner, T., Roth, P. M. and Bischof, H. (2013). Robust real-time tracking of multiple objects by volumetric mass densities, *IEEE Proc. CVPR*.

Raid, D. (1979). An algorithm for tracking multiple targets, *IEEE Trans. Automatic Control* **24**(6): 843–854.

Shabat, G., Shmueli, Y., Bermanis, A. and Averbuch, A. (2015). Accelerating particle filter using randomized multiscale and fast multipole type methods, *IEEE Trans. PAMI* **37**(7): 1396–1407.

Song, S. and Xiao, J. (2013). Tracking revisited using rgbd camera: Unified benchmark and baselines, *IEEE Proc. ICCV*.

Wang, Q., Chen, F., Xu, W. and Yang, M.-H. (2015). Object tracking with joint optimization of representation and classification, *IEEE Trans. Circuits Syst. Video Techn.* **25**(4): 638–650.

Wu, Y., Lim, J. and Yang, M. (2015). Object tracking benchmark, *IEEE Trans. PAMI* **37**(9): 1834–1848.

Zhang, K., Liu, Q., Wu, Y. and Yang, M.-H. (2016). Robust visual tracking via convolutional networks without training, *IEEE Trans. Image Proc.* **25**(4): 1779–1792.

Chapter 5

Trustability Measures of Tracking Algorithms

5.1 Introduction

Quantifying the trustability of a tracking algorithm is a crucial issue for its real-life application. Different tracking algorithms could be effective in different scenarios to different degrees, and may not be even effective at all somewhere. Therefore quantifying the performance of a certain algorithm in a particular environment even for the same task is a necessity. This type of quantification helps the user to chose the most suitable solution with respect to an application. In other words, such quantification would definitely enhance the trustability or reliability of any video tracking-based AI system.

The issue of quantification of the performance of any tracking algorithm has been addressed for more than a decade. There exist several indices to evaluate the performance of tracking (Black *et al.*, 2003; Kasturi *et al.*, 2009; Pal and Chakraborty, 2013, 2017; Nawaz *et al.*, 2014; Čehovin *et al.*, 2016; Fang *et al.*, 2017). The existing indices are of various kinds and different indices fit with different requirements. For example, some of them focus on single object tracking, some focus on multiple object tracking and a few focus on video quality degradation during tracking. The characteristic differences among the working principles of different types of metrices for tracking evaluation are described as follows.

- Multiple Tracking Evaluation (Nawaz *et al.*, 2014): The correctness in object ID (identity) throughout the sequence is primarily measured while evaluating any multiple object tracking algorithm.

That is, a certain object should be identified as that same object throughout the sequence of a tracking algorithm. The other metrics like shift in tracker position have less priority in designing of a multi-tracking evaluation index.

- Single Tracking Evaluation (Čehovin *et al.*, 2016): The problem of single object tracking can be said to be a special case of multiple object tracking. But its performance evaluation includes lots of other features. The performances of tracking algorithms under different challenges like various lighting conditions, shape/size change of object, camera motion are considered during the evaluation. The location of the tracker is given the highest priority while quantifying the performance of an algorithm.
- Evaluation with Quality Degradation (Fang *et al.*, 2017): All the frames in a video sequence may not be of same quality even if captured by the same sensor. Quality degradation in the sequence may take place. There exist a few measures to quantify the performance of tracking over gradually degraded video sequences.

Apart from the aforementioned grouping, the measures can be classified into two categories. Most of the indices work with a comparison to the ground truth information. The underlying assumptions of these measures are that the annotated ground truths of the datasets are always available. There exist a few measures which work with estimated trajectory of moving objects. Annotated ground truth information is not required there. Those measures work well when occlusion or overlapping occurs in the video sequence.

Here in this chapter our objectives are: (i) to describe a few quantitative measures which work without ground truth information and (ii) to experimentally demonstrate the trustability of different existing benchmark tracking methods, including the ones described in the earlier chapters, under different challenging scenarios with the evaluation indices described here. The remaining sections of this chapter are organized as follows. A study over different tracking evaluation metrices which work with ground truth information is provided in Section 5.2. The working principles of the indices which work without ground truth information are described in Section 5.3. The effectiveness of these indices in identifying mis-tracked frames is shown in Section 5.4. The trustability of different benchmark tracking methods under various scenarios are quantified with different indices in Section 5.5. The overall conclusion of this chapter is drawn in Section 5.6.

5.2 Different Performance Evaluation Metrics

- OTE (Black *et al.*, 2003) computes the average positional distance between ground truth and the estimated track. The assignment associates an estimated track with the ground truth track that minimizes the average Euclidean distance across their common frames.
- TRDR (Black *et al.*, 2003) quantifies the overall performance in a frame with the ratio of the number of correctly-tracked targets (true positives) and the number of ground truth targets. An estimation is considered to be a true positive if the centroid of the ground truth bounding box lies (coincides) within the estimated bounding box. If no centroid of ground truth bounding boxes lies in an estimated bounding box, then the estimation is considered to be false positive.
- FAR (Black *et al.*, 2003) measures the tracking performance in some frame in terms of the ratio between the number of incorrectly-tracked targets (false positives) and the total number of targets.
- TDR (Black *et al.*, 2003) evaluates the tracking performance throughout the sequence as the ratio between the number of true positive targets in the estimated tracker and the number of frames where the corresponding ground truth exists within the tracker.
- MOTP (Kasturi *et al.*, 2009) is a spatio-temporal measure that computes the amount of overlap between the estimated and ground truth tracks.
- MODA (Kasturi *et al.*, 2009) computes the tracking performance in each frame individually by combining the information about the number of false negative estimations and the number of false positive estimations.
- IDC (Yin *et al.*, 2007) counts the number of identity (ID) changes in multi-target tracking corresponding to all ground truth tracks. Each of the estimated bounding boxes in every frame is assigned to the ground truth bounding box with an overlap larger than a predefined threshold. When the overlap between an estimated track and the ground truth track falls below the threshold, an ID change is considered to have occurred.

- METE (Nawaz *et al.*, 2014) is an overlap-based measure and it combines the accuracy and cardinality error. It computes the mismatch between the estimated and ground truth states at a certain frame. It also evaluates the changes in the size of extended targets.
- MELT (Nawaz *et al.*, 2014) evaluates the tracking accuracy across the sequence in a parameter-independent manner and enables analysis at different levels of accuracy. It detects whether a ground truth track is associated with more than one estimated track due to ID changes at track level.

Apart from the aforementioned ground truth based indices, the details of a few more quantitative indices, as defined in Pal and Chakraborty (2013, 2017), for tracking are provided in the next section (Section 5.3). These indices neither require any ground truth information nor any trajectory estimation. The object models in the previous frame(s) are taken into account and the similarities of those to the tracked region in the current frame are measured in case of these indices. Sometimes the estimated object model may also be used while defining the indices. Five such indices are described here. Three out of them were formulated considering RGB feature values (Pal and Chakraborty, 2013). The other two (Pal and Chakraborty, 2017) incorporated depth feature into account for accuracy measure. Note that all these indices measure the accuracy of the tracked region(s) and not the object ID. However, the change in object ID between consecutive frames will automatically be reflected since the RGB and depth features in such cases will vary between the object model and the tracked region.

5.3 Indices Without Ground Truth Information

5.3.1 *RGB-based Evaluation Indices*

Three measures (k-index, f-index and b-index) are described here to evaluate the tracking performance (Pal and Chakraborty, 2013). The merits of rough set theory are explored while defining the k-index. The ratio of the number of features present in the tracked region to that of the object model is used while defining the f-index. Bhattacharya Distance (Comaniciu *et al.*, 2003) with some modifications is used in the formulation of b-index.

5.3.1.1 *k-index*

As mentioned in Section 1.4.3 of Chapter 1, we know that $POS_P(Q)$ signifies the part of the data in Q that could be classified employing only the knowledge of P. The same concept was applied while formulating *k-index* in Pal and Chakraborty (2013). During the formulation of the *k*-index for tracking evaluation, the following information was considered, viz., the target object model (T_Mod) and the region of interest (ROI_t) in the current frame. Since the classification of the moving object (ROI_t) in the current frame is carried out by employing the knowledge of the object model (T_Mod) received from the previous frame, the value of $POS_{T\_Mod}(ROI_t)$ is expected to provide the tracked region. It is assumed that the moving object models in the frames have several common features, throughout the sequence, and there will be very less change in target models between two consecutive frames as only a few features are likely to be affected. So, the measure of how much of the knowledge of ROI_t has been evaluated from knowledge T_Mod was considered to be a good measure to evaluate the tracking performance. In rough set theory, the aforesaid concept is defined as follows:

$$k = \gamma_{T\_Mod}(ROI_t) = \frac{|POS_{T\_Mod}(ROI_t)|}{|U|} \tag{5.1}$$

where $|.|$ denotes the cardinality of the set. The higher the value of k, the more is the dependency. If $k = 1$, the entire knowledge of (ROI_t) has been evaluated by employing the knowledge T_Mod.

Here, the union of T_Mod and ROI_t has been considered as the universe of classes (U). The k-value reflects how properly the T_Mod can reconstruct the target model in the current frame, thus it evaluates the performance of T_Mod. That is, if there is any fault in selection of T_Mod, then $|POS_{T\_Mod}(ROI_t)|$ will be small and $|U|$ will be large; thereby giving smaller value of k in the current or tth frame. So, if a frame has smaller k-value, then it can be said that the frame is mis-tracked or over-tracked or under-tracked. In other words, the k-measure can automatically determine if the target model has been reconstructed properly in the current frame using the information of the previous frame.

5.3.1.2 *f-index*

Let a feature present in the region of interest in current frame (ROI_t) be represented as M and the set of features similar or closer to it in the target model (T_Mod) be denoted by N. Let the distance between them be expressed as

$$d = Dist(M, N) \tag{5.2}$$

where, $Dist(.)$ is the distance function used to compute the distance between M and N. The f-measure is computed as the ratio of the summation of the feature distance and the total number of features in the ROI_t. That is,

$$f = \frac{\sum_{i=1}^{F} Dist_i}{F} \tag{5.3}$$

where, F is the total number of features in the target model T_Mod and $Dist$ is the same as in Eq. 5.2. It is expected that all the pixels of the object model in the previous frame will be present in the current frame with a certain shift in the current frame. Then, the feature distance will be zero for each individual feature. Henceforth, it can be concluded that, if the summation of feature distance is small, then the value of f will also be small. That means, there exist more pixels within the ROI_t that are similar to those in T_Mod. So, lower value of f for a frame indicates better tracking in the frame. Any frame having a value of f greater than a certain threshold may be said to be mis-tracked or over-tracked (the tracker contains more redundant features) or under-tracked (the tracker does not contain sufficient features to define the equivalence classes).

5.3.1.3 *b-index*

The color distributions of the target model and the reconstructed object are considered in this measure. Here, a target is represented based on the distributions of the levels in the target model and the tracked region. Let the total range of levels be divided into m number of bins/segments. The size of a bin is dependent on the number of occurrences of the pixels of the levels within that bin; thereby making it wider for levels with higher number of occurrences.

In case of video tracking, the object model region and tracked region are supposed to contain pixels with similar levels. It can be said that the levels where the occurrences of pixels are maximum in both the target model and tracked region are similar. Further, the probability of deviation in the levels with such maximum occurrence of pixels would be higher than those for the levels with less occurrence of pixels.

Let p_u be the number of occurrences of pixels in the uth bin and be represented as follows:

$$p_u = \sum_{i=1}^{n} \delta[bin(x_i) - u]$$

where δ is the Kronecker delta function, $bin(x_i)$ denotes the bin where the ith pixel is and n is the total number of pixels in the region. Then, the sizes of the bins are defined in terms of the following ratio:

$$u_i : u_2 : .. : u_j : .. : u_m = p_1 : p_2 : .. : p_j : .. : p_m. \tag{5.4}$$

The target model is represented as

$$p_{u_j} = \sum_{i=1}^{n} \delta[bin(x_i) - u_j]. \tag{5.5}$$

Let q_{u_j} characterize the tracked regions corresponding to the u_j^{th} bin. Then Bhattacharya Distance between p_{u_j} and q_{u_j} is

$$b = 1 - \sum_{j=1}^{m} \sqrt{(p_{u_j} q_{u_j})}. \tag{5.6}$$

As b represents the distance between two distributions, lower value of b reflects better tracking.

It may be mentioned that a method for object tracking based on Bhattacharya Distance was explained by Comaniciu *et al.* (2003). However, unlike the one defined for b-index, the object was represented there with bins of equal size. Figure 5.1 demonstrates the significance of unequal-size bins over equal bins as a tracking measure. Examples of a target model region and tracked region are shown in Fig. 5.1(a). Their representations with normalized histograms are shown in Fig. 5.1(b), where the dark blue and cyan colored plots show the normalized histograms of the target model and tracked regions respectively with equal bins, whereas the red (target model) and green (tracked region) plots show the same with

(a)

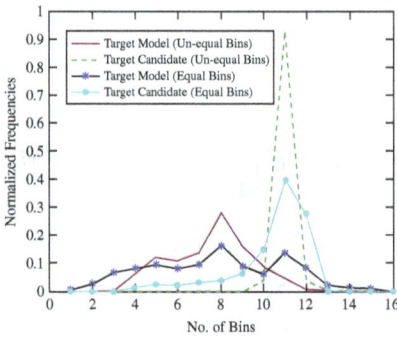 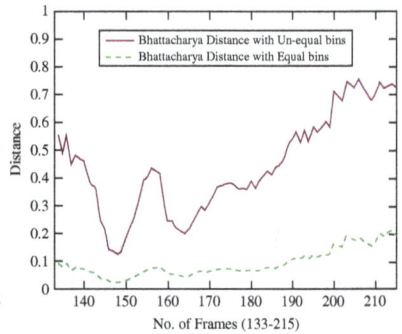

(b)   (c)

Fig. 5.1: (a) Target model and tracked region, (b) their normalized histogram-based representation with equal and unequal bins and (c) Bhattacharya Distance obtained with equal (green) and unequal (red) bins over frame nos. 134–215 of the *Surveillance Scenario Sequence* from PETS-2000.

unequal bins. From the aforesaid plots it is seen that the consideration of unequal histogram bins results in more clear separability between the target model and the target candidate (tracked region). Further, as an example, the Bhattacharya Distance obtained over the tracking results of Frame nos. 134–215 from the *Surveillance Scenario Sequence*, PETS-2000 is shown in Fig. 5.1(c), where the green line shows the Bhattacharya Distance with bins of equal size and the red line represents the same with bins of unequal size. Here too, the Bhattacharya Distance, as computed for b-index with equal-sized bins provides clear peaks and valleys; thereby making it easier to find out the mis-tracked frames.

5.3.2 *Tracking Indices With Depth Feature*

In Section 1.2.1 (Chapter 1), it is mentioned that a video data can be acquired by color (RGB) as well as depth (D) sensors (Chakraborty and Pal, 2015). The depth values are acquired with Kinect or IR sensors. The moving objects in a sequence are expected to have distinct values in this D-feature space as they move in front of their corresponding background, i.e., nearer to the sensor, if they are not occluded. The IR sensor works based on the reflected light with IR wavelength. So, objects closer to the sensor produce brighter regions. But, in case of Kinect sensors, the objects nearer to the sensors have less depth values resulting in darker regions. These characteristics are used in formulating the indices for evaluating the quality of tracking. Two such indices defined by Pal and Chakraborty (2017) involving both Kinect and IR sensed data, and 3-D neighborhood granules (Section 3.3.1) are explained here. The foreground segments of the current and previous frames are the inputs required for the evaluation of FS in the current frame.

5.3.2.1 E_D *Index: Foreground Segmentation*

This index incorporates the edge information to measure the accuracy in the tracking of the foreground segment (FS). In case of IR sensor, the moving object region is brighter resulting in higher gray level, which implies that if a neighborhood of an edge pixel has brighter or same gray level, then the pixel will belong to the object; otherwise the background. The situation will be the reverse in case of Kinect sensor.

For a given foreground segmented region FS in a frame, obtained using IR sensor, the false positive (FP_{ep}) and false negative (FN_{ep}) values for an edge pixel E_{ep} with a window of $[x_i], i \in w \times w$ around it are computed as

$$FP_{ep} = |\{ep_i : i \in w \times w \ and \ ep_i < E_{ep}, \ if \ ep_i \in FS\}| \quad (5.7a)$$

$$FN_{ep} = |\{ep_i : i \in w \times w \ and \ ep_i \geq E_{ep} \ if \ ep_i \notin FS\}| \quad (5.7b)$$

where $|.|$ represents the cardinality of the set. The E_D index of the foreground segment (FS) incorporating the values obtained from Eq. (5.7) is defined as

$$E_D(FS) = \frac{\sum_{ep \in EP} \frac{FN_{ep} + FP_{ep}}{w \times w}}{EP} \qquad (5.8)$$

where EP is the total number of neighborhood edge pixels present in FS.

If instead of IR sensor, Kinect sensor is used, then it maps the depth in the reverse way, i.e., the edge-neighborhood pixels within moving segments have same or lower values than that of E_{ep}. In this way, Eq. (5.7) defining FP and FN gets changed with the conditions $ep_i > Eep_i$ and $ep_i \leq E_{ep}$, and E_D index is calculated accordingly. Lower value of this index implies higher accuracy of the foreground segmented sequence FS, i.e., better tracking of FS.

5.3.2.2 $\aleph_I$ *Index: Mis-classification*

This index is computed with the information of the 3-D spatio-temporal neighborhood granules used for tracking. Its expression is given exploiting the characteristics of IR sensor. That is, the fore-ground segment (FS) would always have brighter values than its background implying that the third dimension (representing different values) of the 3-D object granules should always contain positive values.

As we are dealing with neighborhood granules, the maximum deviations in the values of the points within a granule can be Thr_t (Eq. 3.4). But the amount of deviation may differ within a set (object/background) if it contains several granules. This deviation should not be much in ideal cases until there occur some mis-classifications. This index $\aleph_I$ reflects those cases. Let the object(s) (O_I) and background (B_I) sets over $\aleph_{sp-tmp}$ be

$$O_I = \{\tau_{med}(\aleph_{O_i}) \quad \forall \, \aleph_{O_i} \in FS\} \qquad (5.9a)$$

$$B_I = \{\tau_{med}(\aleph_{B_i}) \quad \forall \, \aleph_{B_i} \notin FS\} \qquad (5.9b)$$

where $\aleph_{O_i}$ and $\aleph_{B_i}$ represent the ith object and background granules, respectively, and τ_{med} is computed according to Eq. (3.3). The amount of scattering (in terms of statistical mean and maximum deviation) in each of the object (d_o) or background (d_b) sets is then computed as follows:

$$d_o = \frac{mean(O_I)}{maxdev(O_I)} \tag{5.10a}$$

$$d_b = \frac{mean(B_I)}{maxdev(B_I)}. \tag{5.10b}$$

In Eq. (5.10), d_o is supposed to be higher as the object granules have higher values, and d_b lower. Therefore, lower value of d_o indicates that some part(s) of the background may get included in the object set, resulting in the case of over-tracking. On the other hand, higher d_b indicates incorporation of object granules in background, i.e., under-tracking. The $\aleph_I$ index, defined accordingly incorporating these phenomena, is

$$\aleph_I = \frac{d_b}{d_o}. \tag{5.11}$$

Therefore, the higher the values of $\aleph_I$, the less accurate the tracking.

5.4 Characteristic Differences Among the Unsupervised Measures

The characteristic difference among the aforesaid three RGB-based indices are as follows. k-index measures the knowledge dependency between the target model and tracked region. The resulting decision on tracking using k-index may be wrong if the size of the object changes rapidly. For example, if the object appears suddenly much bigger than it was in the previous frame, then without having the whole object as the tracked region, or having the same-sized tracked region as it was there in the previous frame, the k-value may become higher, which should not be the case. f-index determines the pixel-wise similarity between these two regions considering the neighbor-hood effect. This measure may also fail similarly if the background contains similar attributes. b-index is based on the similarity between color histograms of the two regions. If there is a huge change in color between object models of two frames, then correct tracking may get detected as mis-tracked by this measure. Though the three indices of tracking have some limitations individually, they may be used together judiciously to supplement each other. For example, the sudden change in object size can be detected by b-index, while the huge change in color can be detected by k-index.

The depth-based indices are able to overcome the limitations caused by the RGB valued ambiguities. The characteristic differences between the two depth-based indices are that the formation of E_D (Eq. 5.8) deals only with the accuracy of foreground segmentation, i.e., how accurately the foreground is segmented. On the other hand, $\aleph_I$ deals with the mis-classification of both foreground and background. Note that the limitation of all of the indices without ground truth is that none of them is able to measure the trustability of any tracking algorithm if some occlusion or overlapping occurs there in the sequence.

5.4.1   *Effectiveness of RGB-based Measures:*
Evaluation of Mis-tracked Frames

Here some experimental results (Pal and Chakraborty, 2013) to show effectiveness of RGB-based measures are presented. The k, f and b indices (Section 5.3.1) can detect the over-tracked or under-tracked frames. According to them, lower value of k or higher value of f and b denotes less accurate tracking. We have shown some mis-tracked frames according to all the three measures. Frame nos. 249, 326, 385, 415 from PETS-2001 (2001) and Frame nos. 75, 84, 125, 176 from PETS-2004 (2004) which are detected as 'mis-tracked' are shown in Fig. 5.2 as two examples.

For instance, in Fig. 5.2 (a), the frames where more background areas are covered either horizontally or vertically by the tracker are detected as mis-tracked frames.

In Fig. 5.2 (b), the frames detected as mis-tracked are those frames where huge portion of the background is included in the tracker, or the object (person) itself is not being entirely covered by the tracker. So, both the cases of under-tracking and over-tracking are seen to be detected by the measures.

One may note that, all of the measures are dependent upon the object model in the previous frame. So, if the target gets mis-tracked in any one of the frames, these measures can detect, but if the tracker is continuously mis-located to some other object, these measures may fail.

(a)

(b)

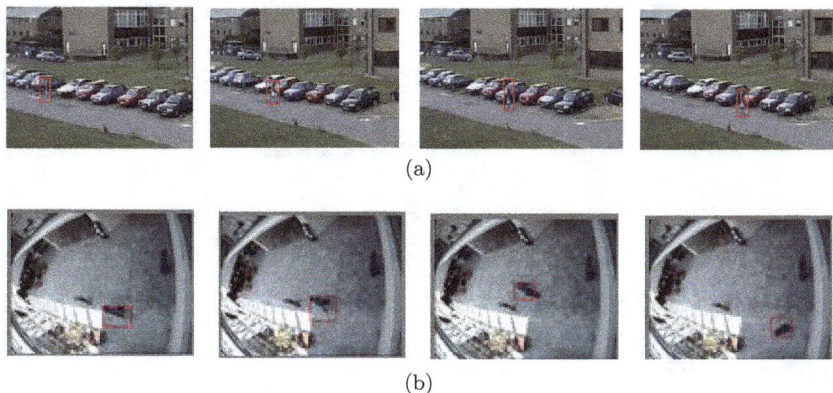

Fig. 5.2: Mis-tracked frames (a) nos. 249, 326, 385, 415 from $P-01$-sequence and (b) 75, 84, 125, 176 from $Walk-03$-sequence.

(a)   (b)

Fig. 5.3: Frame no. 1299 from the *Baggage Detection Sequence* from AVSS-2007. (a) Mis-tracked frame and (b) corrected.

Let us consider Fig. 5.3 as another example of tracking in a sequence. It can be seen in Fig. 5.3(a) that the frame was initially mis-tracked. The values of k, f and b were 0.984 (high), 0.2353 (low), 0.3625 (not low), respectively. As we know, high k-value and low f- and b-value reflect better tracking, here the b index reflects the mis-tracking and hence it gets corrected automatically (Pal and Chakraborty, 2013). The tracking result after correction is shown in Fig. 5.3(b) where b value is reduced to 0.208. In this way, every mis-tracked frame gets corrected in the process of tracking and one can have a satisfactory tracking result throughout all the sequences.

(a)

(b)

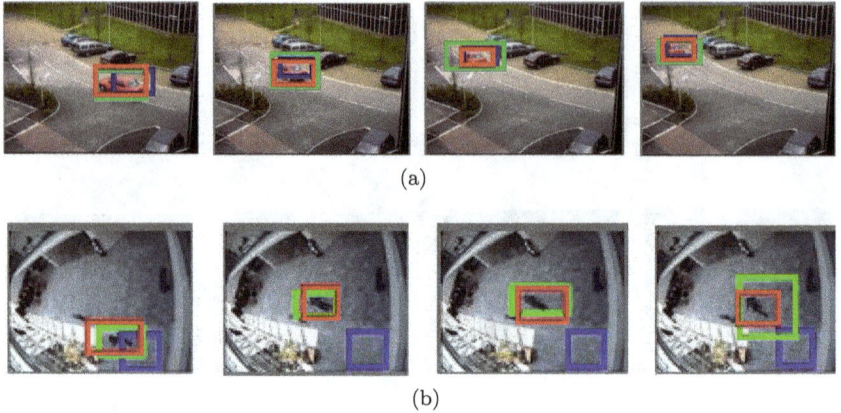

Fig. 5.4: Tracking result: (a) Frame nos. 134–215 from the *Surveillance Scenario sequence* from PETS-2000 and (b) Frame nos. 50–132 from the *Walk3 sequence* from PETS-2004. Trackers with Red, Green and Blue correspond to RE-SpTmp, MoG and MS, respectively.

Now we provide a comparative performance between three methods to demonstrate the suitability of these indices. Methods considered are RE-SpTmp (Chapter 2) and two popular existing methods, viz, Mean Shift tracking (MS) (Comaniciu *et al.*, 2002) and Mixture of Gaussian-based background adaptation (MoG) (Stauffer and Grimson, 1999a). The comparisons are made visually as well as quantitatively with the k, f and b measures for two sequences (*Surveillance Scenario Sequence* from PETS-2000, the *Walk3 sequence* from PETS-2004), as an example. The visual comparative results are shown in Fig. 5.4. The values of k, f and b obtained according to Eqs. (5.1, 5.3, 5.6) for the two sequences applying the three defined methods (viz. RE-SpTmp, mean-shift and MoG) are shown in Figs. 5.5–5.7, respectively. One more numerical comparison is done based on the centroid distance, i.e., how far the tracker centroid is from the ground truth object centroid. Euclidian distance metric is used here. This is a popular measure to evaluate the tracking performance. In Fig. 5.8, values of the centroid distance for the said two sequences corresponding to the three methods are shown. As seen from these figures, the RE-SpTmp method results in higher value of k, and lower values of centroid distance, f and b; thereby signifying more accurate tracking in the former than the other two methods.

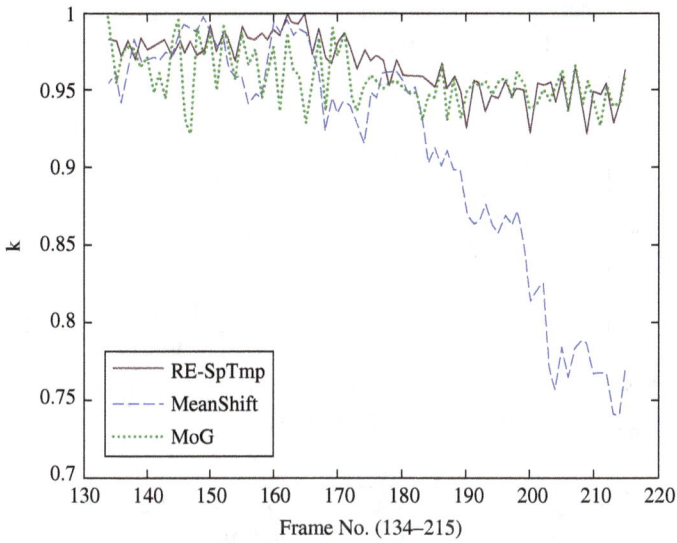

(a)

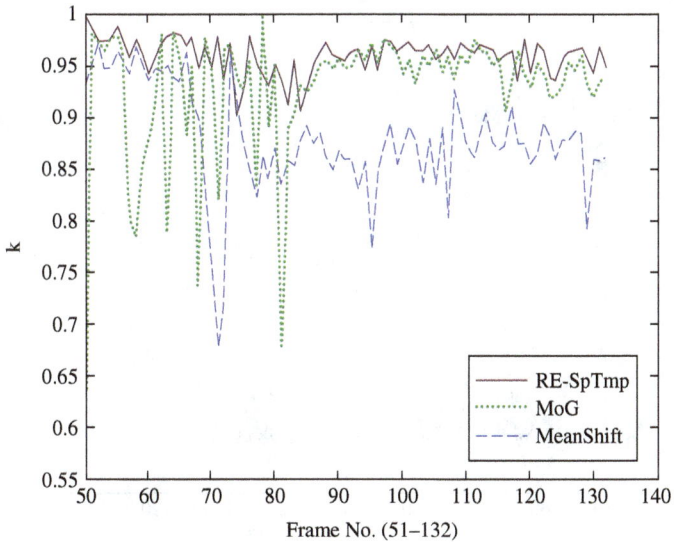

(b)

Fig. 5.5: Values of k obtained for: (a) Frame nos. 134–215 from the *Surveillance Scenario sequence* from PETS-2000 and (b) Frame nos. 50–132 from the *Walk3 sequence* from PETS-2004.

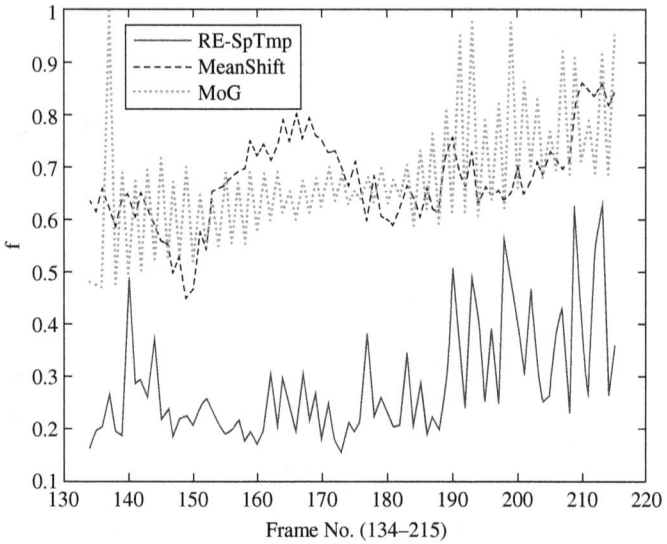

(a)

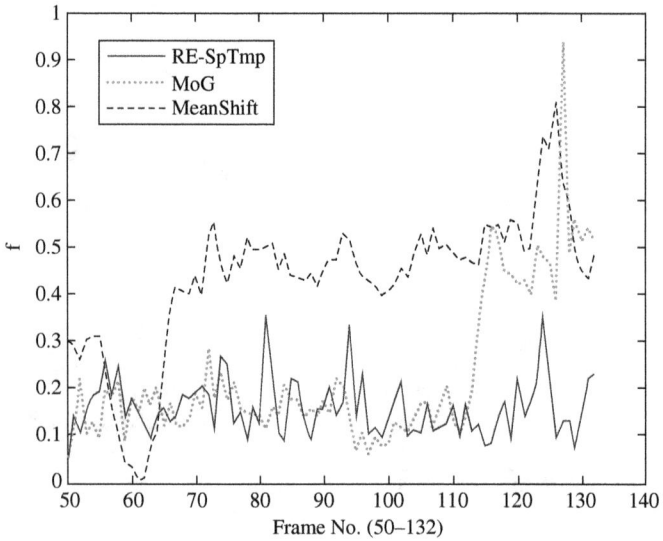

(b)

Fig. 5.6:   Values of f obtained for: (a) Frame nos. 134–215 from the *Surveillance Scenario sequence* from PETS-2000 and (b) Frame nos. 50–132 from the *Walk3 sequence* from PETS-2004.

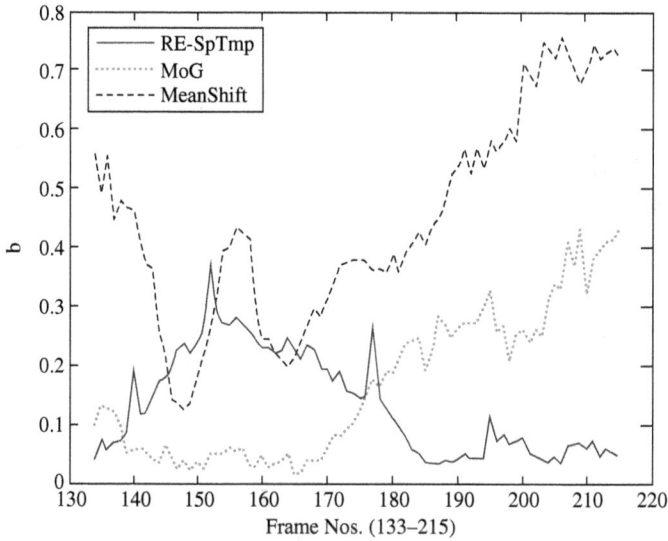

(a)

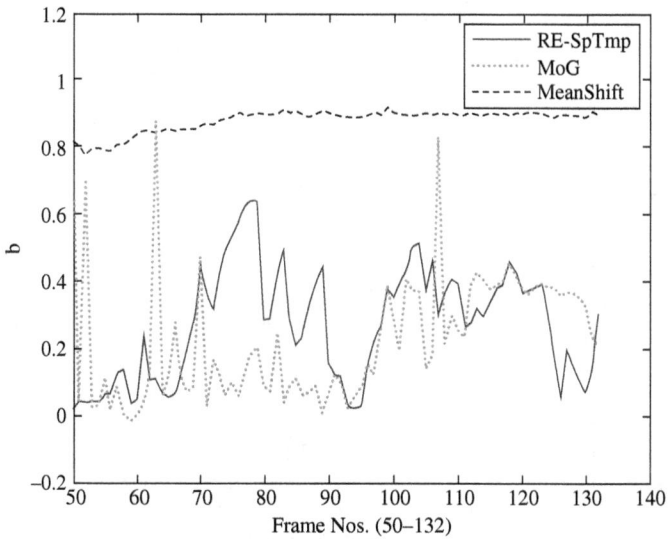

(b)

Fig. 5.7: Values of b obtained for: (a) Frame nos. 134–215 from the *Surveillance Scenario sequence* from PETS-2000 and (b) Frame nos. 50–132 from the *Walk3 sequence* from PETS-2004.

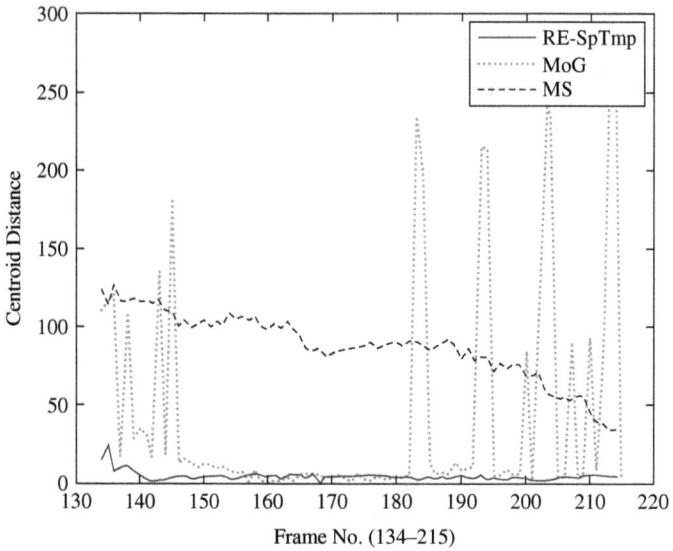

(a)

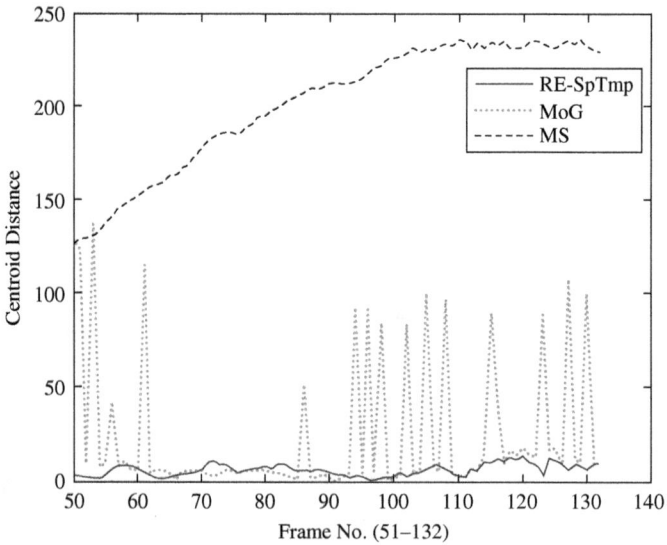

(b)

Fig. 5.8:   Values of *Centroid Distance* obtained for: (a) Frame nos. 134–215 from the *Surveillance Scenario sequence* from PETS-2000 and (b) Frame nos. 50–132 from the *Walk3 sequence* from PETS-2004.

For example, it can be seen from Fig. 5.4(a) that in case of *Surveillance Sequence* the meanshift (MS) method provides the worst tracking result, MoG performs better than MS and the Re-SpTmp method provides the most accurate result among the three methods. The visual results are reflected in quantitative measures also. The ground truth-based measure depicted in Fig. 5.8(a) shows the same. In Fig. 5.5(a), it can be noticed that the average value of k is the lowest for meanshift (MS) method, higher than MS, but lower than the RE-SpTmp one for MoG and the highest for the RE-SpTmp. From Fig. 5.6(a), the f-values, as expected, are lower for RE-SpTmp method and higher for meanshift and MoG. The average b-values are the lowest for the RE-SpTmp method, higher than the RE-SpTmp, but lower than MS for MoG and the highest for MS, and these are shown in Fig. 5.7(a). Table 5.1 depicts a ranking where the columns denote the best (Rank1) to worst (Rank3) methods among the three, as decided according to the average values of k, f and b indices in the respective rows. The RE-SpTmp method is seen to be the best by all the indices for both the sequences, except one (b-index for *Walk3*). So, it can be said that the aforementioned RGB-based measures (k, f and b) are effective to evaluate the performance of tracking quantitatively.

5.4.2 *Validation of Depth-based Measures*

The effectiveness of depth-based indices $\aleph_I$ (Eq. (5.11)) and E_D (Eq. (5.8)) is shown along with that of the popular ground

Table 5.1: Ranking of methods according to k-index, f-index and b-index for *Surveillance Scenario*, and *Walk3* sequences.

Surveillance Scenario	Rank1	Rank2	Rank3
k-index	$RE - SpTmp$	MoG	MS
f-index	$RE - SpTmp$	MS	MoG
b-index	$RE - SpTmp$	MoG	MS
Walk3 Sequence			
k-index	$RE - SpTmp$	MoG	MS
f-index	$RE - SpTmp$	MoG	MS
b-index	MoG	$RE - SpTmp$	MS

truth-dependent indices, namely, centroid distance, false positive pixels (FP) and false negative pixels (FN) over six methods. These methods are SPG (Pan and Schonfeld, 2011), PLS-tracking (Wang *et al.*, 2012), NRBFG (Pal and Chakraborty, 2017), CEMT (Milan *et al.*, 2014), LLAS (Bai *et al.*, 2015) and GMOT (Khan and Gu, 2013) as described in Chapter 3. The CD value for every frame is shown here, whereas for FP and FN, average values are given. The values of these indices obtained for tracking of M_9 sequence with the six methods SPG (blue, '.$-$'), PLS (green, '*$-$'), NRBFG (red, '+$-$') ,CEMT (yellow, 'o$-$'), LLAS (magenta,'$\triangle-$'), and GMOT (black, '> $-$') are shown graphically in Fig. 5.9. The comparative visual results for Frame nos. 13, 22, 32, 39 of M_9 are shown in Fig. 3.11.1 as examples. In case of Frame no. 13, (Fig. 3.11.1(a)), the performances for PLS, NRBFG, LLAS and CEMT are almost equally good. These are well reflected by $\aleph_I$ and E_D indices (Fig. 5.9(a) and 5.9(b)). Similar nature can be seen with the corresponding values of CD. In case of Frame no. 22 (Fig. 3.11.1(b)), SPG results in over-tracking by including some part of the head which does not belong to the object, and hence increases the intra-granular deviations in the object set resulting in higher $\aleph_I$ index as well as higher value of CD. The inclusion of a part of the head leads to increase in FP_{ep} and FN_{ep} values (Eq. (5.7)), and the E_D index. In the cases of the next two frames, 32 and 39 (Fig. 3.11.1(c) and (d)), PLS cannot detect one of the moving objects, which results in higher intra-granular deviation in background set, and high $\aleph_I$ index (CD is also high for this frame). The performance of the other two methods is satisfactory for these two frames, though SPG cannot cover all the objects resulting in higher $\aleph_I$ and E_D values (same is reflected by CD) compared to that of LLAS, NRBFG and CEMT. E_D index can not quantify the mis-tracking by PLS as it takes care only of the fore-ground segments and their corresponding background. E_D and $\aleph_D$ indices for these two frames show that less over-tracking or under-tracking occurs in the foreground segments of SPG compared to those of PLS, NRBFG, LLAS, GMOT and CEMT, i.e., the foreground part which gets segmented is more correct for SPG.

The average values (in terms of %) of false positive (FP) pixels (the background pixels classified as the foreground) and false negative (FN) pixels (foreground pixels classified as background) for this sequence (M_9) are shown in Fig. 5.9(d) and 5.9(e), respectively.

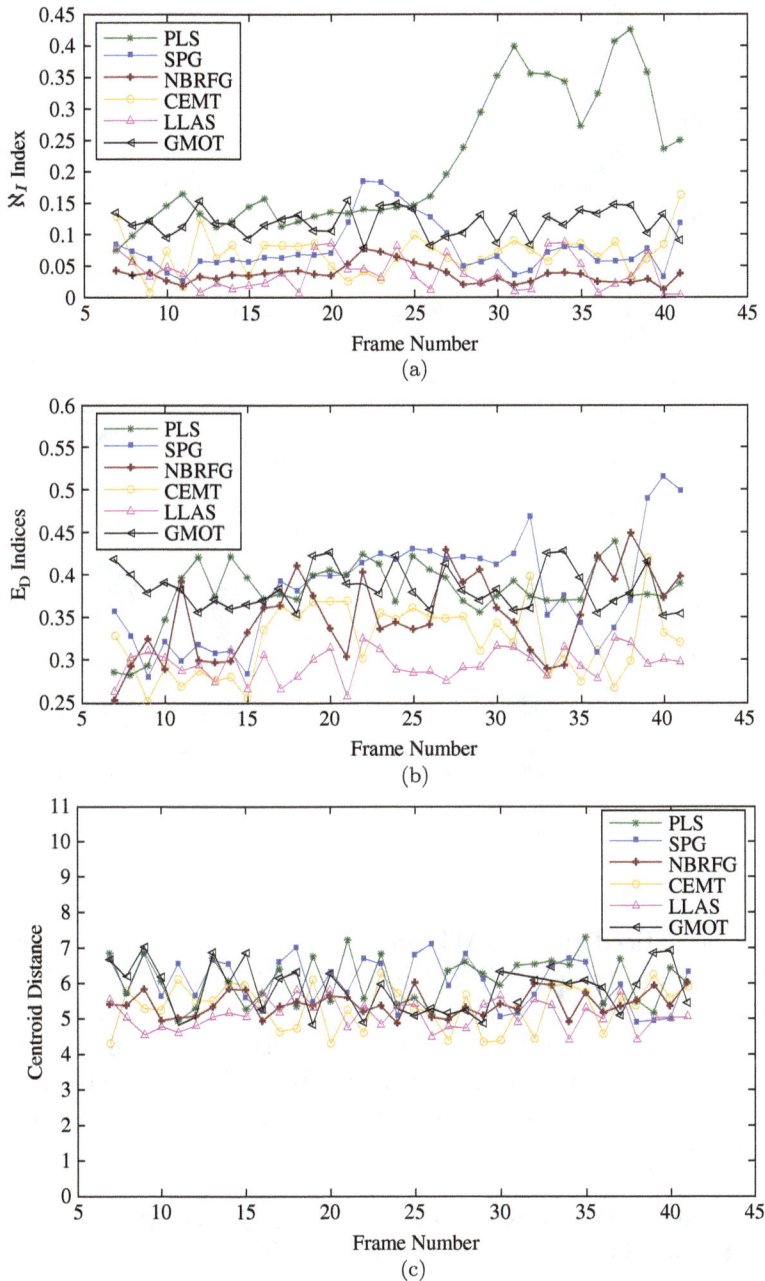

Fig. 5.9: Quantitative comparisons over M_9 Sequence using (a) $\aleph_I$ Indices, (b) E_D Indices, (c) Centroid distance (CD), (d) average % of false positive pixels and (e) average % of false negative pixels.

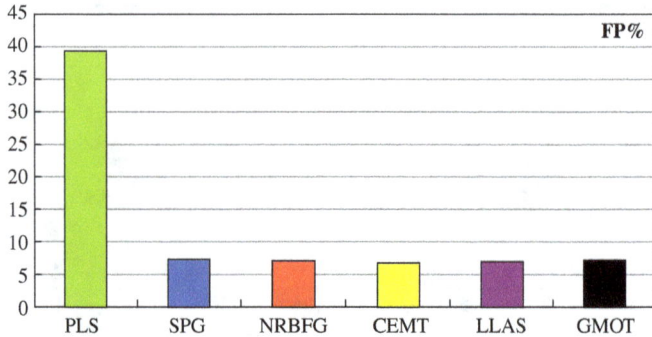

(d)

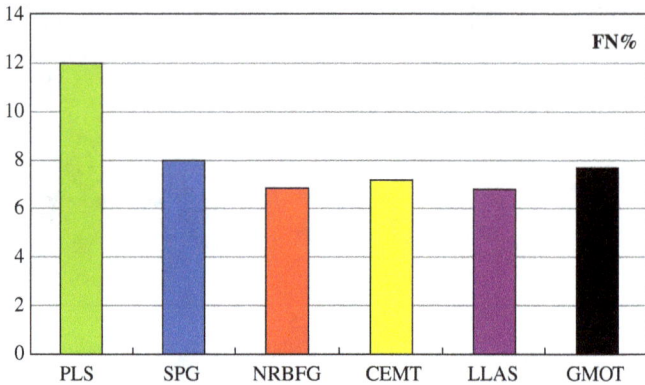

(e)

Fig. 5.9:   (*Continued*)

It can be noticed that CEMT provides the least FP value, whereas LLAS provides the least FN value. The NBRFG provides satisfactory results in terms of both. PLS gives high FN values compared to the other methods. This is expected, as this method fails to detect another moving hand in the sequence M_9. This mis-classification was also reflected by $\aleph_I$. These demonstrate the effectiveness of the E_D and $\aleph_I$ measures.

5.5   Trustability Measure of Different Methods Under Different Scenarios

Different types of efficient measures of tracking performance have been discussed in this chapter so far. Let us now explain the trustability for different benchmark tracking algorithms, including those described in Chapters 2, 3 and 4 under different scenarios. Here the datasets considered involve some of those which were reflected in the earlier results. The scenarios considered are: (i) single object tracking with single still camera, (ii) multiple object tracking with single still camera and (iii) occlusion/overlapping handling.

The tracking methods considered in the study are: (i) RE-SpTmp (described in Chapter 2), (ii) Mean Shift (Comaniciu *et al.*, 2002), (iii) MoG (Stauffer and Grimson, 1999b), (iv) NRBFG (described in Chapter 3), (v) OM (Milan *et al.*, 2014), (vi) PLS (Wang *et al.*, 2012), (vii) CEMT (Pan and Schonfeld, 2011), (viii) LLAS (Bai *et al.*, 2015), (ix) GMOT (Khan and Gu, 2013), (x) MCIO (Park *et al.*, 2015), (xi) ALIEN (Pernici and Bimbo, 2014), (xii) NRFT (described in Chapter 4), (xiii) DeepTrack (Wang *et al.*, 2015) and (xiv) CNT (Zhang *et al.*, 2014). The trustability of these methods over different scenarios are quantified with different measures in the following sections.

5.5.1 *Trustability for Single Object Tracking*

Here the performance of different methods is quantified for the task of single object tracking. The datasets used are $PETS - 01$, $Walk3$ and M_1. As described earlier, all these sequences contain single object motion with shape/size variation and illumination change. The quantitative measures used for evaluation are: (i) OTE (Black *et al.*, 2003), (ii) MOTP (Kasturi *et al.*, 2009) (described in Section 5.2), (iii) k-index, (iv) b-index and (iv) f-index. The average values of each index found out over the entire sequences of $PETS-01$, $Walk3$ and M_1 are listed here in Tables 5.2, 5.3 and 5.4, respectively, for each of the methods. Evaluations with respect to $\aleph_I$ and E_D indices are also performed here since depth information is available along with the RGB.

One may note that the three videos over which the statistical studies are performed have different characteristics. $Pets01$ contains a single object movement with gradual change in shape, size and direction under uniform illumination. All of the methods under consideration produced satisfactory indices for this video as shown in Table 5.2. That is all of the methods are trustable in such a near to ideal condition. The $Walk3$ sequence contains an illumination change and one

Table 5.2: Performance for $PETS-01$ sequence.

Methods	$RE-SpTmp$	MS	MoG	NRBFG	OM	PLS	CEMT	LLAS	GMOT	MCIO	ALIEN	NRFT	DeepTrack	CNT
OTE	23	61	40	22	23	20	26	18	26	24	32	29	19	21
MOTP	92%	71%	82%	85%	95%	91%	93%	85%	83%	88%	91%	96%	97%	94%
k-index	0.93	0.7	0.73	0.81	0.92	0.91	0.95	0.87	0.92	0.86	0.91	0.94	0.96	0.92
f-index	0.21	0.39	0.32	0.22	0.23	0.20	0.26	0.18	0.24	0.23	0.26	0.27	0.17	0.19
b-index	0.18	0.52	0.32	0.24	0.16	0.19	0.21	0.15	0.21	0.24	0.20	0.15	0.13	0.17

Table 5.3: Performance for *Walk3* sequence.

Methods	RE – SpTmp	MS	MoG	NRBFG	OM	PLS	CEMT	LLAS	GMOT	MCIO	ALIEN	NRFT	DeepTrack	CNT
OTE	11	115	20	12	13	10	16	8	16	14	22	19	9	12
MOTP	93%	28%	86%	88%	96%	93%	95%	88%	86%	91%	94%	95%	98%	94%
k-index	0.94	0.65	0.83	0.91	0.93	0.92	0.96	0.88	0.94	0.89	0.91	0.95	0.97	0.93
f-index	0.11	0.49	0.27	0.15	0.13	0.10	0.16	0.18	0.14	0.13	0.16	0.17	0.07	0.09
b-index	0.18	0.52	0.32	0.24	0.16	0.19	0.21	0.15	0.21	0.24	0.20	0.15	0.13	0.17

Table 5.4: Performance for $M.01$ sequence.

Methods	$RE-SpTmp$	MS	MoG	$NRBFG$	OM	PLS	$CEMT$	$LLAS$	$GMOT$	$MCIO$	$ALIEN$	$NRFT$	$DeepTrack$	CNT
OTE	8	45	38	5	7	9	6	7	8	7	12	9	5	6
$MOTP$	94%	39%	55%	86%	95%	91%	92%	90%	89%	92%	93%	94%	97%	95%
k-index	0.91	0.71	0.73	0.89	0.93	0.94	0.95	0.91	0.92	0.87	0.90	0.94	0.96	0.92
f-index	0.14	0.31	0.29	0.16	0.13	0.11	0.15	0.19	0.15	0.12	0.16	0.18	0.09	0.11
b-index	0.14	0.42	0.39	0.21	0.14	0.15	0.19	0.18	0.21	0.20	0.17	0.14	0.10	0.15
$\aleph_I$-index	0.21	0.66	0.58	0.11	0.14	0.32	0.10	0.08	0.12	0.20	0.17	0.09	0.06	0.08
E_I-index	0.28	0.66	0.54	0.32	0.33	0.36	0.27	0.29	0.39	0.28	0.26	0.29	0.24	0.25

can see that the performance indices availed by MS method deviate here with respect to those of the other methods listed in Table 5.3. That is, the trustability of MS is low if there is an abrupt change in illumination. M_1 contains abrupt change both in motion and direction of the moving object in the sequence. The quantitative indices for the methods MS and MoG show the deviated values (given in Table 5.4) and therefore lower the trustability of these two methods under abrupt change of direction and speed of the moving object.

5.5.2 *Trustability for Multiple Object Tracking*

The efficient indices for different methods for the task of multiple object tracking are explained here. The methods that are considered are: (i) NRBFG (described in Chapter 3), (ii) OM (Milan *et al.*, 2014), (iii) PLS (Wang *et al.*, 2012), (iv) CEMT (Pan and Schonfeld, 2011), (v) LLAS (Bai *et al.*, 2015), (vi) GMOT (Khan and Gu, 2013), (vii) MCIO (Park *et al.*, 2015), (viii) ALIEN (Pernici and Bimbo, 2014), (ix) NRFT (described in Chapter 4), (x) DeepTrack (Wang *et al.*, 2015) and (xi) CNT (Zhang *et al.*, 2014). The indices that are chosen here for the performance evaluation are: (i) TRDR (Black *et al.*, 2003), (ii) FAR (Black *et al.*, 2003) and (iii) IDC (Yin *et al.*, 2007) (described in Section 5.2) along with the depth feature-based indices (iv) $\aleph_I$ and (v) $E - I$.

Here the study is performed over two video sequences with different characteristics. Sequence $2 - b$ contains the movement of several persons in a scene, while in sequence M_4 one hand (object) of a lady initially moves and gets stopped, and the other hand (another similar looking object) starts to move then. The tracking performance for the aforementioned scenarios is verified quantitatively in Tables 5.5 and 5.6. It can be observed form Table 5.5 that all the methods perform well in case of tracking multiple object(s) present in a sequence. There is no ID change for the methods $ALIEN$, $NRFT$, $DeepTrack$ and CNT. That is, the trustability for these methods for correct tracking of multiple objects is quite high. On the other hand, it can be observed from Table 5.6 that the performance of PLS method is quite poor. It is also discussed earlier in Chapter 3 that PLS could not always detect the newly appeared object in a sequence and that is the case here too. The performance of OM method is better than PLS in case of tracking newly appearing object(s), but poorer with respect to the other methods. That is, the trustability of

Table 5.5: Performance for 2b sequence.

Methods	NRBFG	OM	PLS	CEMT	LLAS	GMOT	MCIO	ALIEN	NRFT	DeepTrack	CNT
TRDR	0.79	0.84	0.78	0.81	0.75	0.78	0.84	0.91	0.88	0.85	0.90
FAR	0.21	0.16	0.22	0.19	0.25	0.22	0.16	0.09	0.12	0.15	0.10
IDC	0.08	0.04	0.1	0.08	0.04	0.04	0.02	0.00	0.00	0.00	0.00
$\aleph_I$- index	0.25	0.28	0.30	0.26	0.19	0.26	0.29	0.21	0.22	0.23	0.22
E_I- index	0.22	0.19	0.20	0.23	0.21	0.20	0.19	0.20	0.18	0.17	0.19

Table 5.6: Performance for M_4 sequence.

Methods	NRBFG	OM	PLS	CEMT	LLAS	GMOT	MCIO	ALIEN	NRFT	DeepTrack	CNT
TRDR	0.81	0.68	0.48	0.79	0.82	0.78	0.77	0.89	0.88	0.82	0.91
FAR	0.19	0.32	0.52	0.21	0.18	0.22	0.23	0.11	0.12	0.18	0.09
IDC	0.12	0.30	0.54	0.1	0.1	0.08	0.1	0.12	0.1	0.1	0.1
$\aleph_I$- index	0.35	0.48	0.61	0.36	0.29	0.28	0.32	0.25	0.33	0.31	0.29
E_I- index	0.32	0.49	0.65	0.33	0.31	0.30	0.29	0.28	0.28	0.26	0.30

PLS is quite low for the task of tracking newly appearing objects in a scene.

5.5.3 *Trustability for Tracking with Occlusion/Overlapping*

Here the trustability of different methods is quantified for the task of tracking in presence of occlusion/overlapping. The study involves only those methods (out of the ones mentioned in earlier sections) which are capable of dealing with occlusion/overlapping. These are (i) NRBFG (described in Chapter 3), (ii) OM (Milan *et al.*, 2014), (iii) CEMT (Pan and Schonfeld, 2011), (iv) GMOT (Khan and Gu, 2013), (v) MCIO (Park *et al.*, 2015), (vi) ALIEN (Pernici and Bimbo, 2014), (vii) NRFT (described in Chapter 4) (viii) Deep-Track (Wang *et al.*, 2015) and (ix) CNT (Zhang *et al.*, 2014). The indices used for performance evaluation are: (i) OTE (Black *et al.*, 2003), (ii) MODA (Kasturi *et al.*, 2009), (iii) METE (Nawaz *et al.*, 2014). Two different types of sequence are used. Sequence $A-07$ contains entrance of two individual object(s) one by one, who then get totally occluded to the background, whereas two individual objects get overlapped to each other with different degrees in the sequence $much - 132$.

It is seen from the values of the quantitative indices in Tables 5.7 and 5.8 that the methods like $NBRFG$, $CEMT$ and $GMOT$ have

Table 5.7:   Performance for $A - 07$ sequence.

Methods	$NRBFG$	OM	$CEMT$	$GMOT$	$MCIO$	$ALIEN$	$NRFT$	$DeepTrack$	CNT
OTE	42.2	7.9	31.2	25.2	7.8	8.1	3.1	3.9	7.8
$MODA$	56%	78%	51%	60%	79%	78%	87%	85%	83%
$METE(\mu)$	0.59	0.35	0.57	0.38	0.40	0.42	0.34	0.33	0.34

Table 5.8:   Performance for $much - 132$ sequence.

Methods	$NRBFG$	OM	$CEMT$	$GMOT$	$MCIO$	$ALIEN$	$NRFT$	$DeepTrack$	CNT
OTE	13.4	3	9.2	7.4	2.4	3.2	2.13	1.68	1.91
$MODA$	84%	87%	81%	90%	89%	88%	97%	95%	93%
$METE(\mu)$	0.39	0.25	0.37	0.28	0.30	0.22	0.24	0.23	0.24

low trustability in the task of tracking, if total occlusion to background of the object(s) takes place in the scene. Whereas these three methods are almost equally trustable to the other methods for the task of tracking even in the presence of partial object-to-object overlapping.

5.6 Conclusions and Discussions

In this chapter, we have discussed different indices and methods to evaluate the performance of tracking. These include a few indices which will work in absence of ground truth. Their use in determining the trustability of different benchmark tracking methods over different scenarios is demonstrated. It can be concluded that the same method cannot remain equally trustable if the scenario gets changed. Therefore, the study in this chapter will definitely help the users to chose the proper method depending upon the requirement.

So far we have discussed different techniques and measures related to the task of tracking. In the next two chapters we are going to address two different applications of video processing. The problem of object recognition from a scene will be discussed in the next chapter and the problem of recognition of an event in a video sequence will be addressed in Chapter 7.

Bibliography

Bai, T., Li, Y.-F. and Zhou, X. (2015). Learning local appearances with sparse representation for robust and fast visual tracking, *IEEE Trans. Cyberns.* **45**(4): 663–675.

Black, J., Ellis, T. and Rosin, P. (2003). A novel method for video tracking performance evaluation, *Proc. Joint IEEE Int. Workshop on VS-PETS*, pp. 125–132.

Chakraborty, D. B. and Pal, S. K. (2015). Neighborhood granules and rough rule base in tracking, *Nat. Comput. Springer (Special Issue on Pattern Recognition and Mining)* **15**(3): 359–370.

Comaniciu, D., Ramesh, V. and Meer, P. (2002). Mean shift: A robust approach towards feature space analysis., *IEEE Trans. PAMI* **24**: 603–619.

Comaniciu, D., Ramesh, V. and Meer, P. (2003). Kernel-based object tracking, *IEEE Trans. PAMI* **25**(5): 564–575.

Čehovin L., Leonardis, A. and Kristan, M. (2016). Visual object tracking performance measures revisited, *IEEE Trans. Image Proc.* **25**: 1261–1274.

Fang, Y., Yuan, Y., Li, L., Wu, J., Lin, W. and Li, Z. (2017). Performance evaluation of visual tracking algorithms on video sequences with quality degradation, *IEEE Access* **5**: 2430–2441.

Kasturi, R., Goldgof, D., Soundararajan, P., Manohar, V., Garofolo, J. and Bowers, R. (2009). Framework for performance evaluation of face, text, and vehicle detection and tracking in video: Data, metrics, and protocol, *IEEE Trans. PAMI* **31**(2): 319–336.

Khan, Z. and Gu, I.-H. (2013). Nonlinear dynamic model for visual object tracking on grassmann manifolds with partial occlusion handling, *IEEE Trans. Cyberns.* **43**(6): 2005–2019.

Milan, A., Roth, S. and Schindler, K. (2014). Continuous energy minimization for multitarget tracking, *IEEE Trans. PAMI* **36**(1): 58–72.

Nawaz, T., Poiesi, F. and Cavallaro, A. (2014). Measures of effective video tracking, *IEEE Trans. Image Proc.* **23**(1): 5–43.

Pal, S. K. and Chakraborty, D. (2013). Unsupervised tracking, roughness and quantitative indices, *Fundamenta Infrmaticae* (*IOS Press*) **124**(1–2): 63–90.

Pal, S. K. and Chakraborty, D. B. (2017). Granular flow graph, adaptive rough rule generation and tracking, *IEEE Trans. Cyberns.* **47**(12): 4096–4107.

Pan, P. and Schonfeld, D. (2011). Video tracking based on sequential particle filtering on graphs, *IEEE Trans. Image Proc.* **20**(6): 1641–1651.

Park, C., Woehl, T. J., Evans, J. E. and Browning, N. D. (2015). Minimum cost multi-way data association for optimizing multitarget tracking of interacting objects, *IEEE Trans. PAMI* **37**(3): 611–624.

Pernici, F. and Bimbo, A. (2014). Object tracking by oversampling local features, *IEEE Trans. PAMI* **36**(12): 2538–2551.

PETS-2001 (2001). *IEEE Int. WS Perfor. Evaluation of Tracking and Surveillance.*

PETS-2004 (2004). *IEEE Int. WS Perfor. Evaluation of Tracking and Surveillance and EC Funded CAVIAR project/IST 2001.*

Stauffer, C. and Grimson, W. E. L. (1999a). Adaptive background mixture models for real-time tracking, *IEEE CVPR*, pp. 246–252.

Stauffer, C. and Grimson, W. E. L. (1999b). Adaptive background mixture models for real-time tracking, *Computer Vision and Pattern Recognition, 1999. IEEE Computer Society Conference on*, Vol. 2, IEEE, pp. 246–252.

Wang, Q., Chen, F., Xu, W. and Yang, M.-H. (2012). Object tracking via partial least squares analysis, *IEEE Trans. Image Proc.* **21**(10): 4454–4465.

Wang, Q., Chen, F., Xu, W. and Yang, M.-H. (2015). Object tracking with joint optimization of representation and classification, *IEEE Trans. Circ. Syst. Video Technol.* **25**(4): 638–650.

Yin, F., Makris, D. and Velastin, S. A. (2007). Performance evaluation of object tracking algorithms, *Proc. Joint IEEE Int. Workshop on VS-PETS*, pp. 1–8.

Zhang, K., Nanjing, C., Zhang, L. and Yang, M.-H. (2014). Fast compressive tracking, *IEEE Trans. PAMI* **36**(10): 2002–2015.

Chapter 6

Object Recognition and Deep Learning

6.1 Introduction

We have so far discussed about different challenges associated with the task of video tracking and how different features of rough set theoretic granular computing could be effective in handling those issues. Here in this chapter we focus on another application of video processing, namely, object recognition from video scene with deep learning. Deep learning-based approaches have now become quite popular for the task of object recognition from images. For the past decade, deep learning (i.e., in-depth learning) has emerged as a new area of machine learning. In deep learning architecture, this problem of object recognition can be regarded as a task of labeling different objects in an image with the correct class as well as predicting the bounding boxes along with probability. Different structures of deep neural networks on this problem have been developed (Erhan *et al.*, 2014; Ren *et al.*, 2015; Redmon *et al.*, 2016; Liu *et al.*, 2016). A neural network (NN) is made up of several neurons. Input neurons get triggered from the environment while the other neurons get triggered through weighted links from previously active neurons. In order to extract the complex representation from rich sensory inputs, human information processing mechanisms suggest the need of deep architectures (LeCun *et al.*, 2015). However, deep learning (DL) often requires hundreds or thousands of images for better performance, unlike the conventional (shallow) learning. The term 'shallow' is meant in contrast to deep (Pal *et al.*, 2020). Therefore,

DL is computationally intensive and requires a high-performance GPU (Graphical Processing Unit). Here comes the relevance of granular computing (GrC) in DL framework to deal with this issue. The present chapter focuses on this, granulated deep learning, for object recognition. Before we discuss the scope of the chapter, we explain the characteristics of deep learning (DL) as a machine learning tool and its relevance, deep architectures, some applications of DL to object recognition and limitations. This is followed by the significance of incorporating the concept of granulation in DL framework, a methodology, recently developed, demonstrating how to integrate these two paradigms for object recognition, and quantification of the performance in linguistic terms.

Machine learning (ML), a branch of artificial intelligence (AI), basically means learning patterns from examples or sample data. Here the machine is given access to the data and has the ability to learn from it. The data (or examples) could be labeled, unlabeled or their combination. Accordingly, the learning could be supervised, unsupervised or semi-supervised. Artificial neural networks (ANNs) that have the ability to learn the relation between input and output from examples are good candidates for ML. ANNs enjoy the characteristics like adaptivity, speed, robustness/ruggedness and optimality. In the early 2000s, certain breakthroughs in multi-layered perceptron (MLP) neural networks facilitated the advent of deep learning. Deep learning (DL) means learning in depth in different stages (Pal, 2018). DL is thus a specialized form of ML which takes ML to the next level in an advanced form. This is characterized by learning the data representations, in contrast to task-specific algorithms.

Deep Learning algorithms/networks are inspired by the structure and function of the human nervous system, where a complex network of interconnected computation units (nodes) works in a coordinated fashion to process complex information. In order to extract the complex representation from rich sensory inputs, human information processing mechanisms suggest the need for deep (learning) architectures (LeCun *et al.*, 2015). Such an architecture usually involves a cascade of multiple layers of nonlinear processing units for feature extraction and transformation. Each successive layer uses the output from the previous layer as input. In deep learning architecture, the problem of object recognition can be regarded as a task of labeling different objects in an image with the correct class as well as predicting the bounding boxes with a high probability. Different structures of deep neural networks on this problem have been proposed (Erhan *et al.*, 2014; Ren *et al.*, 2015; Jia *et al.*, 2014; Redmon *et al.*, 2016).

Convolutional neural network (CNN, or ConvNet) (Krizhevsky *et al.*, 2012) represents one such deep architecture which is most popular for learning with images and video. Like other neural networks, a CNN is composed of an input layer, an output layer and several hidden layers in between. These layers perform one of the three types of operations, i.e., convolution, pooling, or rectified linear unit (ReLU), on the data. Convolution puts the input images through a set of convolutional filters (Krizhevsky *et al.*, 2012), each of which activates certain features from the images. Pooling [14] simplifies the output by performing nonlinear down sampling, and reducing the number of parameters that the network needs to learn about. Rectified linear unit (ReLU) (LeCun *et al.*, 2015) allows for faster and more effective training by mapping negative values to zero and maintaining positive values. These three operations are repeated over tens or hundreds of layers, with each layer learning to detect different features. CNNs have been used for motion detection and object recognition (Liu *et al.*, 2016; Gundogdu and Alatan, 2018; Erhan *et al.*, 2014; Ren *et al.*, 2015; Redmon *et al.*, 2016).

Deep learning (DL) has dramatically improved the state of the art in object recognition (LeCun *et al.*, 2015), AI among other applications. However, since DL relies on sample data (or previous experience), the learning performance depends on the number of such samples. The larger the number, the more accurate the performance. Today, we have abundant data; so DL has become a meaningful choice. DL often requires hundreds or thousands of images for the best results, unlike the conventional (shallow) learning. Therefore, DL is computationally intensive and difficult to engineer. It requires a high-performance GPU (Graphical Processing Unit). For example, deep learning networks like single shot detector (Liu *et al.*, 2016), Faster-R-CNN (Jia *et al.*, 2014; Ren *et al.*, 2015), CFCF (Gundogdu and Alatan, 2018) provide very fast motion detection and object recognition using GPU.

While deep learning is a computationally intensive process and the granular computing paradigm, on the other hand, leads to gain in computation time, a judicious integration of these two was formulated by Pal *et al.* (2020) to make the deep learning framework efficient in terms of computation time. The granulated deep learning-based object detection and tracking method (GrDLODT) embodies such an attempt, where rough set theoretic granular computing is used in CNN for speedy motion detection and moving object recognition.

Further, quantification of the performance in image/video processing, and of interpretation and understanding of scenes, where subjectivity is of major concern, has always been a challenging issue. Z-numbers, proposed recently by Zadeh (1996) provide a framework to quantify the abstraction of semantic information from natural language statements where subjectivity plays an important role in understanding. In a part of the GrDLODT method, the abstract concept of z-numbers was used in interpreting a scene with certainty in terms of recognition of its constituting objects, in natural language.

The basic block diagram of the GrDLODT method is shown in Fig. 6.1 for motion detection and object recognition with linguistic descriptions. Here various kinds of granulation (Chakraborty *et al.*, 2013; Pal and Chakraborty, 2017) based on spatial and temporal segmentation on image frames are used. Rough set theoretic approximations are made to construct the object and background models. Roughness, in terms of lower and upper approximations, is optimized in the process. Unlike the CNN, where the entire image is scanned pixel-by-pixel in the convolution layer, here scanning is done only over the representative pixel of each granule. This lowers the computation time drastically, compromising little with accuracy. Training a deep neural network (DNN) requires a large number of labeled training data, thereby making the process slow. Therefore, transfer learning was used in GrDLODT to reduce this problem, where one uses weights from already trained models of Convolutional Neural

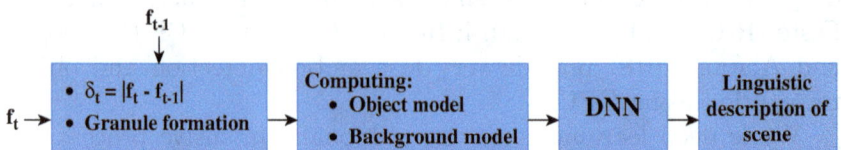

Fig. 6.1: Overall block diagram of the recognition system GrDLOT.

Network (Krizhevsky *et al.*, 2012). GrDLODT method involves recognition of both static objects in the background and moving objects in the foreground separately for scene analysis. The current video frame f_t and one previous frame f_{t-1} are the input to the system for computing the frame difference ($\delta_t = |f_t - f_{t-1}|$) and forming granules. The background model (B_g) is computed only once, while the object models (O_b) are determined for each input frame. B_g and O_b are fed separately into the DNN for recognition of static and moving objects. The output provides a linguistic description of the scene consisting of these objects.

The rest of this chapter is organized in the following way. Section 6.2 describes the characteristics of the granulated deep learning (GDL), after explaining the conventional CNN in brief. This includes the concept, characteristics and relevance of the mechanism of granulated deep learning, and the methodology and algorithms for object tracking and recognition. Results of object tracking and recognition along with comparisons with some state-of-the-art algorithms are described in Section 6.3. Section 6.4.1 describes the Z-number-based measures for object recognition and scene understanding. Section 6.5 provides the conclusions.

6.2   Granulated Deep Learning

6.2.1   *Granulation Techniques*

Here we describe different kinds of granules as used in GrDLODT and their formation, and the method of determining the optimum upper and lower approximations on the granulated image based on rough set theory. All the techniques used in GrDLODT are described in the previous chapters. Here we mentioned them in brief for the convenience of readers.

6.2.1.1   *Image Definition with Rough Sets*

Here the frames of a video sequence are defined as rough sets following the formulation of Section 1.4.7. That is, $\overline{O}_T$ and $\underline{O}_T$ represent the object lower and upper approximations with respect to a gray level threshold T, and $\underline{B}_T$ and $\overline{B}_T$ represent the same for the background set. The object roughness and background roughness are denoted by R_{O_T} and R_{B_T}, respectively.

6.2.1.2 *Formation of Granules*

Different types of granules described in the previous chapters of this book are used here in GrDLODT method. These include: (i) uniform-sized rectangular granules with spatial similarity (Pal *et al.*, 2005), (ii) unequal-sized rectangular granules with gray level and spatial similarities (described in Chapter 2) and (iii) natural arbitrary sized/shaped granules with spatio-color similarity (described in Chapter 3). These are described here in brief for the convenience of the readers.

Equal sized and shaped granules: Let I be an image of size M × N. Let the image be partitioned into a collection of non-overlapping windows of equal size (m × n). Then each window can be considered as a granule.

Unequal sized and regular shaped granules: Here, the granulation is based on quad-tree decomposition of the images. It is described in detail in Section 2.3.1 of Chapter 2.

Arbitrary shaped granules: These are formed by region growing technique based on color similarity among the 8 neighbors of a candidate pixel. The process of forming granules is described in Section 3.3.1 and then the object/background classification is carried out following the rule-base of Table 3.4.3.

Optimum threshold (T_O) for computing lower–upper approximations: After the granulated image plane (I_G) is obtained, different thresholds (T) for object–background separation are considered, and the one which results in minimum roughness is determined for I_G. It is computed similarly as for T^* in Section 2.3.2.

As stated before, the learning with deep neural networks is a slow process. Here we explain how the concept and merits of granular computing can be integrated with a deep neural network CNN to speed up its learning mechanism. The effectiveness of the resulting system (called granulated deep learning network (GrDL)) is demonstrated for motion tracking and scene analysis from videos. Before we explain the features of the GrDL mechanism in GrDLODT, we describe the conventional CNN in brief.

6.2.2 *Conventional Convolution Neural Network*

A CNN has three types of layers, viz, convolution layer, pooling layer and rectified liner unit (ReLU) layer. The convolution layer of deep learning network basically involves shifting of a sliding window all over the image or frame. Let the input to the network be a $32 \times 32 \times 3$ array of RGB (Red–Green–Blue) pixel values. To explain the functionality of a convolution layer, let us imagine a flashlight that is shining over the top left of the image. Let us assume that this flashlight covers a 5×5 pixel area. Let this flashlight slide across all the areas of the input image. The flashlight that covers a 5×5 window in the image may be viewed as a filter. The region of the image that is being shined over is called the receptive field. For mathematical matching, the depth of this filter has to be the same as the depth of the input image, where the intensity of the said flashlight (filter) is representing a 5×5 array of numbers, called weights; so the dimensions of this filter are $5 \times 5 \times 3$ for RGB components. As the filter is sliding (or convolving), around the input image, it is multiplying the values in the filter with the original pixel values of the image. These multiplications are all summed up to get a single number. This process is repeated for every location on the input image. Next step would be moving the filter to the right by S unit, then right again by S, and so on, where S is called the Stride. If S = 1, after sliding the filter over all the locations, one will find out that what is left is a $28 \times 28 \times 1$ array of numbers, which is called an activation map.

After computing the activation map, the next layer is pooling which simplifies the output by down-sampling; thereby reducing the number of parameters in the network. Rectified Linear Unit (ReLU) allows fast training by mapping the negative values to zero and retaining the others as they are. These three operations are repeated several times to obtain a desired (converged) output.

Training a deep neural network from scratch requires enormous labeled (training) data and hence computing power (hundreds of GPU-hours or more). Transfer learning is a technique that shortcuts much of this by taking a piece of a model that has already been trained on a related task and reusing it in a new model (Krizhevsky *et al.*, 2012).

6.2.3 *Granulated Deep Learning: Concepts, Characteristics and Relevance*

The convolution layer of the deep learning network, as described before, basically involves shifting of a sliding window all over the

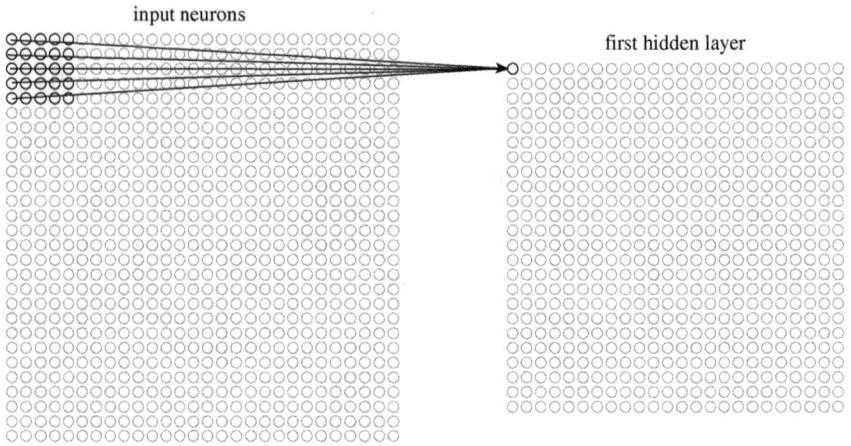

Fig. 6.2:   Activation map.

image or frame for convolving (e.g., multiplication and summation). Here, in the GrDL network of GrDLODT model, instead of using a raw pixel-based frame as input to this convolution layer, that frame is granulated using different approaches, as discussed in Section 6.2.1, before considered as input.

If the original image size is 32 × 32, then after granulation, the image would consist of granules instead of 32 × 32 pixels. The number of such granules is obviously much less than 32 × 32. Let us assume that there are n number of granules, named $g_1, g_2, \ldots, g_n$ in a frame. This granulated frame is used as the input of the first convolution layer. The filter region chooses the top left corner of the image, which belongs to granule g_1, as the receptive field. After the first cell value of the activation map (Fig. 6.2) is computed, the same value is put in the activation map corresponding to the other pixels in g_1. That means, all the remaining pixels which belong to the same granule g_1 were skipped, and one need not compute cell values of the activation map more than once for a particular granule.

Unlike the pixel-based convolution method, where the stride is fixed *a priori*, here the stride automatically selects only the top left corner of each granule as the receptive field; thereby skipping the remaining pixels. By doing so, the selection of appropriate stride (which is crucial in a conventional CNN) does not arise. Note further that, in the pixel-based method, sliding the filter over all the

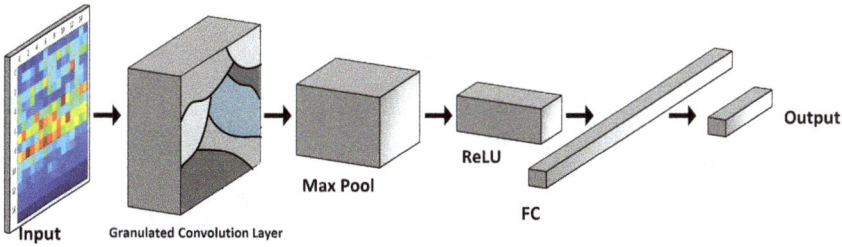

Fig. 6.3:   Layers of granulated deep learning model.

32×32 pixels results in a 28×28 array of numbers. But in GrDLODT method one needs to do the filtering only n times where n $<< 32^2$. Thus the computation time is greatly reduced, although some accuracy may be compromised. Figure 6.3 shows the structure of the GrDL network where the first convolution layer is granulated. Use of this network for object recognition and tracking is described in Section 6.2 where the granulated convolution layer takes the object model and background model as input.

6.2.4   *Granulated Deep Learning: Object Tracking and Recognition*

Given a scene containing static (background) and moving objects, the recognition algorithm has two parts. First, it recognizes the classes to which these objects belong. Secondly, it tracks the objects in motion. In doing so, the method involves classification of static object and moving objects separately by the granulated deep learning model (Fig. 6.3). The entire method of tracking and recognition is explained in brief by a block diagram in Fig. 6.4.

Let f_t and f_{t-1} be the current frame and its immediate previous frame of a video sequence, respectively. Let the frame difference be computed as

$$\delta = |f_t - f_{t-1}|. \tag{6.1}$$

δ, characterizes the change between two consecutive frames, which represents only the moving portion (pixels) in the frames.

The granulation on f_t is formed by the methods described in Section 6.2.1. After granulation, the system computes the upper approximation of the background $\overline{B}_T(= U_B)$ over the granulated

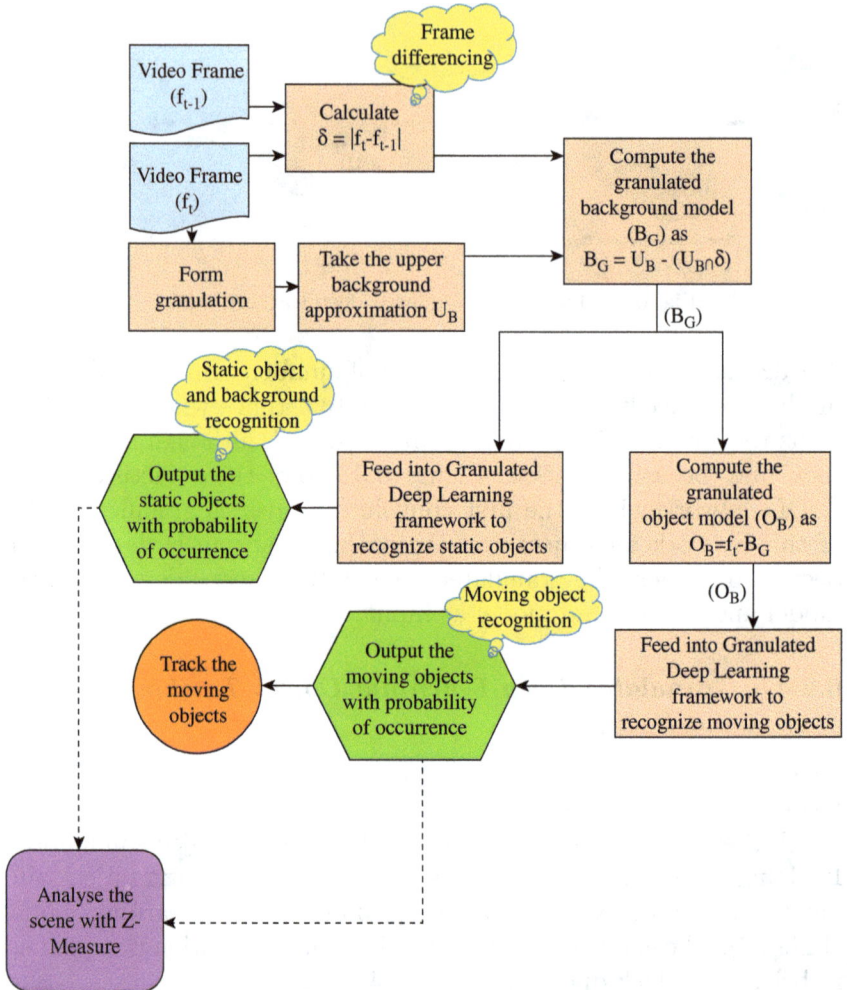

Fig. 6.4: Block diagram of the GrDLODT method for object recognition and tracking.

image, corresponding to the optimum threshold (T_O). That means, U_B is the set of granules where at least one pixel value is less than T_O.

The intersection of upper approximated background (U_B) and frame difference (δ) denotes the portion which erroneously may belong to the background despite being a part of the moving object. Therefore, to ensure that the approximated background (U_B) has no part of moving objects, the intersection of U_B and δ is subtracted from U_B, and this is denoted as the granulated background model (B_G). That is,

$$B_G = U_B - (U_B \cap \delta). \tag{6.2}$$

The granulated object model (O_B) is obtained as,

$$O_B = f_t - B_G. \tag{6.3}$$

Let us now explain the task of recognition and tracking. Here recognition concerns both static (background) and moving objects, and the tracking is done on moving objects. Recognizing static objects (background) is performed only once at the beginning, using the granulated deep learning framework (shown by dashed lines and blocks) with B_G as input. Static objects remain the same throughout the videos. For recognizing moving objects from the input frames f_i ($i = 2, 3, ..., N$, where N is the number of frames), the granulated object model (O_B) of each frame is taken as input to the said granulated deep learning network. Its output denotes the categories of the moving objects, which are then tracked.

The aforesaid steps are explained in Algorithm 4.

The results of recognition of static and moving objects, as obtained in Fig. 6.4, are used for analysis of the input scene. This is done in a new way in terms of linguistic description using Z-numbers (Zadeh, 2011). Such a unique description is provided in Section 6.4.1. Before that, some experimental results on the performance of tracking and recognition of GrDLODT are provided in Section 6.3.

6.3 Experimental Results

The effectiveness of the GrDLODT method in the tasks of (a) motion tracking and (b) object recognition is shown here for three types of granulation techniques. These are: (i) equal-sized and -shaped granules, (ii) unequal rectangular-shaped granules and (iii) arbitrary-shaped granules, into the granulated deep learning framework. Transfer learning by considering weights from the already trained models of CNN was used (Krizhevsky *et al.*, 2012), as described in Section 6.2.2. The trained CNN that is used in GrDLODT method is the Single Shot Detector (SSD) (Liu *et al.*, 2016) with COCO (Common objects in context) dataset (Lin *et al.*, 2014), containing 20 classes of objects (+1 for the background). These objects are airplanes, bicycles, birds,

---

Algorithm 4 Object Recognition and Motion Tracking Using Granulated Deep Learning

---

INPUT: Video frames f_i, $\forall i = 1, 2, ..., n'$ where n' is the number of frames in the input video.

OUTPUT: Frames with recognized objects and bounding box around moving objects.

1: Take the current frame f_t and its immediate previous frame f_{t-1}.

2: Calculate the temporal information $\delta = |f_t - f_{t-1}|$.

3: Perform granulation on the frame f_t.

4: Compute the upper background approximation U_B after the quad-tree decomposition $U_B = \{\bigcup_i G_i, \forall j = 1, 2, ..., mn\}$ such that $P_j < T$ where $P_j \in G_i$ and T is the threshold for distinguishing object and background.

5: Compute the granulated background model $B_G = U_B - (U_B \cap \delta)$.

6: Feed B_G into the granulated deep-learning framework, defined in Section 6.2.4, for static (background) object recognition [Step 6 is a one-time process as the background is fixed for all frames].

7: Compute the granulated object model $O_B = f_t - B_G$.

8: Feed O_B into the granulated deep learning framework, described in Section 3.2, for moving object(s) recognition with probability of occurrence (P_O).

9: Track the moving object(s) by placing bounding boxes around them.

10: Repeat the process for the next frame f_{t+1} onwards.

---

boats, bottles, buses, cars, cats, chairs, cows, dining tables, dogs, horses, motorbikes, people, potted plants, sheep, sofas, trains and tv monitors.

The test dataset consists of seven different types of video sequences: (1) Cam 131 Sequence (Changing appearance scenario) of ICG Lab (Possegger *et al.*, 2013), (2) Girl sequence of TB-50 (*Visual Tracker Benchmark Data*, 2018), (3) PASCAL VOC 2007 (*Visual Tracker Benchmark Data*, 2018), (4) VOT 2017 (Kristan *et al.*, 2016) (Bolt, Gymnastic1 and Godfather sequence), (5) Jurassic Park Intro (*Jurassic Intro Dataset*, 2018), (6) PETS 2009 (Ferryman and Shahrokni, 2009) and (7) Changing Size Car (Patacchiola and

Cangelosi, 2017). The results on some of the frames of each of these video sequences are given here along with the comparison with some state-of-the-art method. These video sequences have both single-type objects and multiple-category objects.

One may note that the granulated deep learning (GrDL) network (Section 6.2) was trained with 20 classes of objects. But the aforesaid seven types of video sequences that were used as test data do not have all those 20 classes in a particular sequence. Some of them contain objects only of a particular class, whereas some others have objects of multiple classes, but not of all categories. The parameter values are dependent on the nature of the input data. For example, the number of previous frames n is dependent on the speed of the video. It was normally chosen as 7 for the sequences with speed of 15 frames per sec. The object–background threshold value T is initially chosen to be 30 as it was found to be experimentally suitable for most of the datasets. If the set $O_T = \phi$, then T was reduced by 5.

6.3.1 *Results of Moving Object Detection and Tracking*

Let us consider the sequences CAM 131 Changing appearance (chap) scenario, PETS 2009, Jurassic Park and Changing Size Car. Both CAM 131 and PETS 2009 have only one type of class, namely person. Datasets Jurassic Park and Changing Size Car, on the other hand, have multiple types of classes, e.g., chair, dining table, person, bottle, airplane and car. Results of tracking and recognition on these sequences are shown in Figs. 6.5 and 6.6 corresponding to objects of single category and multiple categories. Here arbitrary-shaped granules (Section 6.2.1.2) were used in the deep learning network. It is shown that the bounding boxes are decently covering the moving objects in the video frames for both the datasets. Accuracy in recognizing the objects is very high, as depicted therein.

6.3.2 *Comparative Study*

The comparison is done in terms of time and accuracy between the method of granulated deep learning (with 3 × 3, quad-tree decomposition, and arbitrary-shaped granules), deep learning without granulation, and other relevant deep learning algorithms such as CFCF (Gundogdu and Alatan, 2018), CFWCR (He *et al.*, 2017),

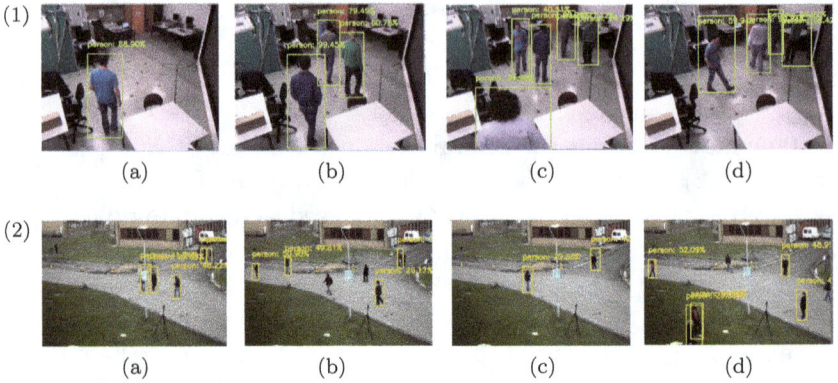

Fig. 6.5: Results of tracking using rectangular granulation by quad-tree-decomposition: Frame nos. 25, 200, 321 and 381 of the Changing appearance (chap) scenario (1) and Frame nos. 105, 194, 424, 530 of the PETS 2009 people tracking dataset (2).

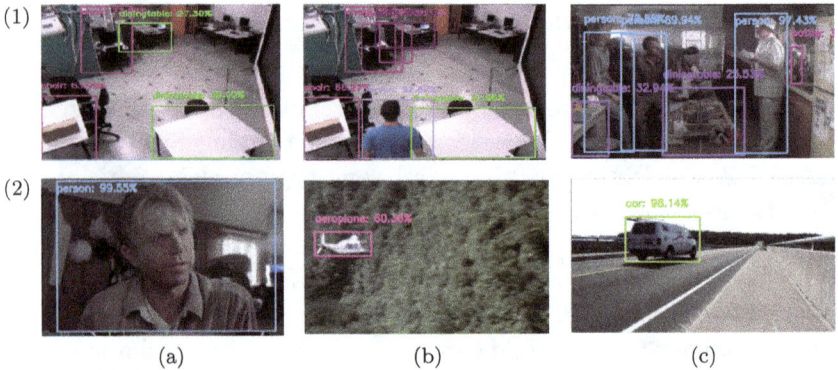

Fig. 6.6: Results of tracking and recognition for multiple categories of objects: 1(a) and 1(b)–Frame nos. 5, 22 of the Changing appearance (chap) scenario, 1(c), 2(a) and 2(b)–Frame nos. 12, 25 and 29 of the Jurassic Park Intro dataset, and 2(c) Frame no. 19 of the Changing Size Car dataset.

LOT (Oron *et al.*, 2015), SCM (Zhong *et al.*, 2014), and SSD (Liu *et al.*, 2016), as applicable to different datasets. Tables 6.1–6.4 demonstrate the comparative results. Tables 6.1 and 6.2 deal with Cam 131 sequence (Changing appearance scenario) and PASCAL VOT 2017 data, respectively, concerning both tracking and recognition of only one type of objects (person). Table 6.3 depicts results

Table 6.1: Time and accuracy comparison for recognition and tracking between GrDLODT (granulated deep learning using 3 × 3 granules, rectangular granules and arbitrary-shaped granules) and deep learning without granulation.

Method	Speed (fps)	Track (%)	Accuracy of detection (%)	Processor
Granulated Deep Learning using 3 × 3 granules	2.2	74.6	62.11	CPU
Granulated Deep Learning using rectangular granules	2	80.1	67.11	CPU
Granulated Deep Learning using arbitrary shaped granules	1.89	81.67	68.56	CPU
Deep learning without granulation	1.6	82.25	70.2	CPU

Table 6.2: Time and accuracy comparison for recognition and tracking between CFCF, CFWCR and GrDLODT (granulated deep learning using 3 × 3 granules, rectangular granules and arbitrary-shaped granules).

Method	Accuracy (%)	Speed (fps)	Processor
CFCF	50.9	1.7	CPU
CFWCR	48.4	1.4	CPU
Granulated Deep Learning using 3 × 3 granules	41.71	2.1	CPU
Granulated Deep Learning using rectangular granules	48.1	1.9	CPU
Granulated Deep Learning using arbitrary shaped granules	48.59	1.5	CPU

Table 6.3: Time and accuracy comparison for tracking between LOT, SCM and GrDLODT (granulated deep learning using 3×3 granules, rectangular granules and arbitrary-shaped granules).

Method	Accuracy (%)	Speed (fps)	Processor
LOT	67.6	0.7	CPU
SCM	69	0.51	CPU
Granulated Deep Learning using 3×3 granules	56.11	1.9	CPU
Granulated Deep Learning using rectangular granules	67.15	1.6	CPU
Granulated Deep Learning using arbitrary-shaped granules	67.88	1.2	CPU

Table 6.4: Time and accuracy for object recognition between SSD and GrDLODT (granulated deep learning using 3×3 granules, rectangular granules and arbitrary-shaped granules).

Method	Accuracy (%)	Speed (fps)	Processor
SSD	74.3	0.5	CPU
Granulated Deep Learning using 3×3 granules	54	2.59	CPU
Granulated Deep Learning using rectangular granules	69	2	CPU
Granulated Deep Learning using arbitrary-shaped granules	72.3	1.87	CPU

for only tracking using Girl sequences of TB-50 data with one type of objects. On the other hand, Table 6.4 deals only with recognition, but it is for multiple classes of objects (viz, cat, dog, train, bird, cup, person, bottle, cow, cycle, dining table, horse, airplane, bus and chairs) using the PASCAL VOC dataset.

For the purpose of fair comparison of GrDLODT method with the state-of-the-art algorithms (viz CFCF (Gundogdu and Alatan, 2018), CFWCR (He *et al.*, 2017), LOT (Oron *et al.*, 2015), SCM

(Zhong *et al.*, 2014) and SSD (Liu *et al.*, 2016)), the same dataset and tasks as used by those authors in their respective studies were used. Accordingly, the comparing deep learning methods are CFCF and CFWCR in Table 6.2 for both tracking and recognition using the data PASCAL VOT 2017; LOT and SCM in Table 6.3 for tracking using the Girl sequence OF TB-50; and SSD and Faster R-CNN in Table 6.4 for object recognition using PASCAL VOC 2007. In these tables, the accuracy was measured based on the distance between the centroids of the ground truth (provided with the data) and the obtained foreground segment from the respective frames. Time was computed in terms of the number of frames processed per second in the Intel core i-5 processor.

As expected, among the various kinds of granulation in the GrD-LODT method, the one with arbitrary-shaped granules, characterizing natural granulation, provides the best performance, while the one with 3×3 granules needs the least computation time to process a frame. This is true for all the cases in Tables 6.1–6.4; thereby demonstrating the effectiveness of this concept of granulated deep learning. The comparing deep learning methods usually provide better accuracy, requiring more computation time, except CFCF (Table 6.2), which performs better in terms of both time and accuracy than methods using arbitrary-shaped granulation. Table 6.2 deals with the dataset PASCAL Visual Object Tracking 2017, consisting of only one type (human) of objects. Here GrDL with arbitrary-shaped granules is superior to CFWCR than that of GrDLODT in terms of both accuracy (e.g., 48.59% vs. 48.4%) and speed (e.g., 1.5 fps vs. 1.4 fps). On the other hand, using the same dataset, CFCF is able to process 1.7 frames per second (fps), whereas it is 1.9 fps by GrDL with rectangular granules. That means GrDL is faster than CFCF, but the accuracy is little less (48.1% vs. 50.9%). GrDL takes less time to process a frame, because instead of scanning every pixel as in CFWCR, it does the operation on granulated image in the granulated convolution layer (Fig. 6.3).

It may be mentioned that the quality of segmentation of objects in a frame is crucial for their tracking. In a part of the investigation in Pal *et al.* (2020), the performance of segmentation of all the methods were compared corresponding to Table 6.1 quantitatively using Beta index (Pal *et al.*, 2000). For a given number of object regions, the higher the value of Beta, the better the segmentation.

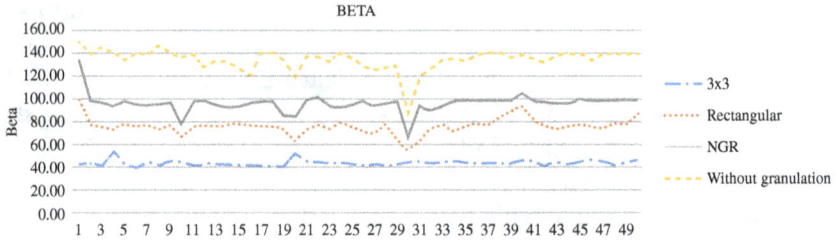

Fig. 6.7: Variation of Beta index for the 'Changing Appearance' sequence.

The variation of Beta over frames for the 'Changing Appearance' sequence is shown in Fig 6.7. As expected, the arbitrary-shaped granules leading to natural granulation provide highest Beta index, as compared to the other two types of granulation. As granulation leads to information loss, the conventional deep learning algorithm, involving no granulation, provides better segmentation in terms of Beta index among all, however at the cost of computation time.

Based on the aforesaid results of recognition of static and moving objects in GrDLODT, the input scene is analyzed in terms of linguistic description using Z-numbers (Zadeh, 2011). This description provides a more natural way of interpreting the scene with certainty for better understanding. Details of the methodology are described in the next section after defining the Z-numbers with an example.

6.4 Z-number-based Scene Description

The Z-numbers (Zadeh, 2011) provide a new fuzzy-set-theoretic approach to Computing With Words (CWW) (Zadeh, 1996). In the CWW paradigm, the perceptions are encoded in the words and phrases used to describe events. This is inspired from the remarkable perception-based decision-making ability of the human brain. The concept of Z-number correlates to the issue of certainty of information. A Z-number has two tuples, $Z = (A, B)$. The first tuple is A, which is a constraint, allowed to take on the values of X (a real-valued uncertain variable, interpreted as the subject of Y). The second tuple, B, is a measure of reliability of the first component. Normally, A and B are described in a natural language, as words or clauses, and are both fuzzy numbers (Banerjee and Pal, 2013).

Example of Z-numbers: Let us consider a statement Y := It takes Jack about 10 minutes to reach school from his house. Then, X := Distance from Jack's house to school, and $Z = \langle$about 10 minutes, usually$\rangle$. Here, A is context-dependent while B summarizes the conclusiveness in the relevance of A given X within the context of Y.

6.4.1 *Formulation of Measures*

In GrDLODT method, Z-numbers are used to define measures for object recognition and scene understanding. As described in Section 6.3, the training dataset has 20 classes. The granulated neural network predicts the class in which the objects of an unknown scene belong with certain probability associated with it. Based on this prediction, some rule-bases were defined (Pal *et al.*, 2020) to provide the information on A and B of Z-numbers.

Here A is defined as the set of classes of similar kinds of objects. That is, A= {Animal, Flying Objects, Transport, Two Wheeler, Furniture}. For example: Animal denotes cat, dog, cow and horse, Flying Object denotes airplane and bird, Transport denotes car, bus, train, Two wheeler denotes cycle and motor bike, and Furniture denotes sofa, chair, dining table. Similarly, A can have another possibility as A= {Indoor, Outdoor} where, Indoor and Outdoor refer to the location of the aforesaid objects.

B is the set of certainty values. For example, B= {Most likely, May be, Not likely at all}. Determining values of B is done in two steps. First, some thresholds on P_O (probability of occurrence) of individual objects were assigned, as obtained from the granulated deep learning network. For example, if P_O of an object is less than 20%, then it is labeled as 'Not Likely' (NL), if the probability is greater than 20% but less than or equal to 70%, then it is labeled as 'May Be' (MB), and for greater than 70% it is 'Most Likely' (ML). Having these linguistic values of individual objects, different rule-bases in the second step were formed in order to determine the values of B corresponding to each element of A.

Table 6.5 shows, for example, the rule-base, thus formed, for the class 'Animal'. If all the initial object labels are NL, then the label of the animal class is NL. If all of them are MB, then result is ML, otherwise if at least one of them is MB, then the result is MB. And if at least one of them is ML, then the result is ML. A few such

Table 6.5:   Rule-base for animal.

cat	dog	cow	horse	ANIMAL
NL	NL	NL	NL	NL
NL	NL	NL	MB	MB
NL	NL	MB	NL	MB
NL	NL	MB	MB	MB
NL	MB	MB	MB	MB
NL	MB	MB	NL	MB
MB	MB	MB	MB	ML
ML	NL	NL	NL	ML
ML	ML	ML	ML	ML

Table 6.6:   Rule-base for flying object.

airplane	bird	FLYING OBJECT
NL	NL	NL
NL	MB	MB
MB	NL	MB
MB	MB	ML
NL	ML	ML
ML	ML	ML

representative rules are listed in Table 6.5 for convenience. Similar rules, derived for Flying Object, Transport, Two Wheeler and Furniture classes, are shown in Tables 6.6, 6.7, 6.8 and 6.9, respectively.

Now for an unknown video frame f_t, its Z-number(s) with A and B, as described before, can be computed to predict the certainty of its constituent objects. Consider, for example, the Frame 22 of the Chap scenario (Possegger *et al.*, 2013) (Section 6.3). As per the output of granulated neural network, the frame contains sofa with 30%, 4 chairs with percentages 66, 38, 59 and 32, one table with 32% accuracy and a person with 92% accuracy. Z-numbers of the Frame f_t are accordingly computed as $Z_O = \langle$Furniture, Most likely$\rangle$ and $Z_O = \langle$Person, Most likely$\rangle$. The results (Z-number(s) for object recognition) from various frames of different sequences like Cam 131 Sequence of ICG Lab (Possegger *et al.*, 2013), Jurassic Park (*Jurassic Intro Dataset*, 2018) and Changing Size Car (Patacchiola and Cangelosi, 2017) are listed in Table 6.10. One may note that, for

Table 6.7:   Rule-base for transport.

car	bus	train	TRANSPORT
NL	NL	NL	NL
NL	MB	NL	MB
MB	NL	MB	MB
MB	MB	MB	ML
MB	MB	ML	ML
NL	NL	ML	ML
ML	ML	ML	ML

Table 6.8:   Rule-base for two wheeler.

cycle	motor bike	TWO WHEELER
NL	NL	NL
NL	MB	MB
MB	NL	MB
MB	MB	ML
NL	ML	ML
ML	ML	ML

Table 6.9:   Rule-base for furniture.

sofa	chair	dining table	FURNITURE
NL	NL	NL	NL
NL	MB	NL	MB
MB	NL	MB	MB
MB	MB	MB	ML
MB	MB	ML	ML
NL	NL	ML	ML
ML	ML	ML	ML

objects which are not appearing in the scene, their Z-numbers would contain certainty value as 'Not likely'. Those cases have not been included in Table 6.10, except the Frame number 560 of Jurassic Park, which is shown for illustration.

Table 6.10: Linguistic description of frames using Z-numbers: Object recognition.

Sequence	Frame no.	Objects@
Chap	22	$Z_O = \langle$Person, Most likely$\rangle$, $Z_O = \langle$Furniture, Most likely$\rangle$
Chap	5	$Z_O = \langle$Furniture, May be$\rangle$
Chap	12	$Z_O = \langle$Furniture, Most likely$\rangle$
Jurassic Park	490	$Z_O = \langle$Person, Most likely$\rangle$
Jurassic Park	498	$Z_O = \langle$Flying Object, May be$\rangle$
Jurassic Park	553	$Z_O = \langle$Flying Object, Most likely$\rangle$
Jurassic Park	554	$Z_O = \langle$Flying Object, May be$\rangle$
Jurassic Park	560	$Z_O = \langle$Flying Object, Not likely$\rangle$, $Z = \langle$Transport, Most likely$\rangle$
Jurassic Park	945	$Z_O = \langle$Person, Most likely$\rangle$, $Z = \langle$ Vehicle, May be$\rangle$
Changing Car Size	19	$Z_O = \langle$Vehicle, Most likely$\rangle$

For scene classification, whether it is indoor or outdoor, the rule-base formed was as shown in Table 6.11. The corresponding Z-numbers of the same set of frames, as used in Table 6.10, with respect to scene classification are depicted in Table 6.12. Consider, as an example, the frame number 22 (Table 6.10), whose Z-measures are computed as $Z_O = \langle$Person, Most likely$\rangle$ and $Z_O = \langle$Furniture, Most likely$\rangle$. According to the rule-base (Table 6.11), if the Furniture class exists in the scene, with a certainty value, then that frame should be classified as Indoor, with the same certainty value. Accordingly, in Table 6.12, the Z-number for scene classification of this frame is $Z_S = \langle$Indoor, Most likely$\rangle$.

6.4.2 *Significance of Z-number-based Measure*

The aforesaid description, based on Z-number, provides granulated information of a scene for its understanding. The linguistic description of a frame, as obtained using Z-numbers, has several

Table 6.11: Rule-base for scene classification.

Animal	Flying object	Vehicle	Animal	Furniture	Indoor	Outdoor
NL	NL	NL	NL	NL	May be	Not likely
NL	NL	NL	MB	NL	Not likely	May be
MB	NL	NL	NL	NL	Not likely	May be
MB	MB	MB	MB	NL	Not likely	Most likely
MB	MB	MB	ML	NL	Not likely	Most likely
ML	ML	ML	ML	NL	Not likely	Most likely
ML	ML	ML	ML	ML	Not likely	May be
NL	NL	NL	NL	MB	May be	Not likely
NL	NL	NL	NL	ML	Most likely	Not likely

Table 6.12: Linguistic description of frames using Z-numbers: Scene classification.

Sequence	Frame no.	Objects
Chap	22	$Z_S = \langle$Indoor, Most likely$\rangle$
Chap	5	$Z_S = \langle$Indoor, May be$\rangle$
Chap	12	$Z_S = \langle$Indoor, Most likely$\rangle$
Jurassic Park	490	$Z_S = \langle$Indoor, May be$\rangle$
Jurassic Park	498	$Z_S = \langle$Outdoor, May be$\rangle$
Jurassic Park	553	$Z_S = \langle$Outdoor, Most likely$\rangle$
Jurassic Park	554	$Z_S = \langle$Outdoor, May be$\rangle$
Jurassic Park	560	$Z_S = \langle$Outdoor, Most likely$\rangle$
Jurassic Park	945	$Z_S = \langle$Outdoor, May be$\rangle$
Changing Car Size	19	$Z_S = \langle$Outdoor, Most likely$\rangle$

applications. For example, from their values over frames, one can notice the sudden appearance or disappearance or occlusion of some object(s) in video sequences. Let us consider Table 6.10 and the Frames 490, 498, 553, 554 and 560. In the Frame 490, there was no flying object, resulting in $Z_O = \langle$Flying Object, Not likely$\rangle$; and in Frame 498, the Z-number was $Z_O = \langle$Flying Object, May Be$\rangle$, i.e., there may be a trace of some flying object. In all the frames up to 553, it was definite that there exists a flying object. In Frame 554, the certainty of having a flying object is 'May be'. Finally in Frame 560, there was no flying object as $Z_O = \langle$Flying Object, Not likely$\rangle$.

From this, one can infer that, the presence, absence or sudden appearance of an object can be noticed automatically using Z-numbers. Accordingly, needful action can be taken, depending on the application, for example, surveillance.

6.5   Conclusions and Discussions

This chapter dealt with the problem of object recognition from video scene using deep learning in the framework of granular computing. Significance of deep learning architecture in this task is explained. The relevance of integrating the concept of granular computing into deep neural network for strengthening the performance of the latter is explained. A recent method of Pal *et al.*, named GrDLODT (Pal *et al.*, 2020), involving such granulated deep learning (GrDL) is discussed in detail illustrating its different features and merits in the aforesaid context. Apart from object recognition, the problem of scene interpretation in linguistic terms is addressed. It is shown how the abstract concept of Z-numbers can be used to quantify the abstraction of semantic information in object classification from a scene; thereby providing a more natural way of interpreting a scene with certainty for its better understanding in natural language.

The GrDLODT uses the theory of rough set in defining the background models over the granulated image planes, as constituted by either 3x3 granules, rectangular granules or unequal-shaped and -sized (arbitrary) granules. The method involves separate recognition of static and moving objects. The trained network that is used in GrDLODT method is the Single Shot Detector (SSD) with COCO dataset containing 20 classes of objects. Test data has 7 sets of benchmark video sequences, containing 14 types of multiple objects, involved in tasks like object recognition and/or tracking. Performance in tracking and recognition, with respect to speed and accuracy, is compared with several state-of-the-art deep learning algorithms, and GrDLODT is proven to be superior.

The conventional convolutional neural network (CNN) for deep learning is expensive in terms of time and resource requirement. Incorporation of the concepts of granular computing in deep learning (i.e., GrDL) reduces the computation time significantly, as it involves scanning only over the granules, instead of each pixel in the input frame. Further, the problem of selecting the appropriate stride, which is crucial in CNN, does not arise in GrDLODT. The method provides a balanced trade-off between speed and accuracy in tracking as compared to pixel level deep learning, and can successfully handle the challenging cases like tracking partial overlapped objects and suddenly appearing objects. Arbitrary-shaped granules, resulting in natural granulation as expected, provide superior performance compared to 3×3 granules and rectangular granules.

The concept of using Z-numbers, in providing a granulated linguistic description of a scene, is unique. Computation of Z-measure involves the granular information of objects and the certainty of their appearance in the video. This provides an information measure of a scene, by consolidating the classification scores of individual objects belonging to a particular category. This measure can be considered as an index for detecting automatically the change in information in a scene, for example, due to occlusion, sudden appearance and disappearance of objects. Therefore it promises to have several real-life applications.

The GrDLODT method basically demonstrates a way of integrating the concept of granular computing with deep learning networks for speedy computation, and using Z-numbers for quantification of scene understanding. This can be viewed just as basic modules. Further generic variants of these modules can be derived for improved performance depending on the application domain and users' need. For example, recently an attempt has been made by Pramanik *et al.* (n.d.) to develop a granulated RCNN (G-RCNN) for object detection which is an advanced version of the well-known Fast RCNN (Girshick, 2015) and Faster RCNN (Ren *et al.*, 2015) for extracting regions of interest (RoIs) by incorporating the unique concept of granulation in a deep CNN. G-RCNN uses videos directly as input, rather than static images. Here granulation with spatio-temporal information enables more accurate extraction of RoIs in unsupervised mode; thereby enhancing the online detection speed and accuracy.

It may be mentioned here about a few assumptions behind the development of GrDLODT, leading to some of its limitations. For example, it was assumed that the varieties of object(s) present in a sequence will be restricted to those of COCO dataset. Accordingly, the test videos used belong to these categories. Further, it was assumed that no initial occlusion/overlapping will be present while defining the object–background sets. However, although the effectiveness of GrDLODT has been demonstrated on simple indoor and outdoor video sequences containing various common objects (e.g., man, car and chair), the model can equally be implemented on larger datasets with same merits and characteristic features.

So far in this book we have discussed the challenges and models related to tracking and object recognition from scene through video processing. In the next chapter we are going to address another very important issue: unsupervised prediction of events from video sequences in the framework of rough set theoretic granular computing.

Bibliography

Banerjee, R. and Pal, S. K. (2013). The z-number enigma: A study through an experiment, *Soft Computing: State of the Art Theory and Novel Applications*, Springer, pp. 71–88.

Chakraborty, D., Shankar, B. U. and Pal, S. K. (2013). Granulation, rough entropy and spatiotemporal moving object detection, *Applied Soft Computing* **13**(9): 4001–4009.

Erhan, D., Szegedy, C., Toshev, A. and Anguelov, D. (2014). Scalable object detection using deep neural networks, *Proc. IEEE Conf. Comput. Vis. Pattern Recognit.* pp. 2147–2154.

Ferryman, J. and Shahrokni, A. (2009). Pets2009: Dataset and challenge, *Performance Evaluation of Tracking and Surveillance (PETS-Winter), 2009 Twelfth IEEE International Workshop on*, IEEE, pp. 1–6.

Girshick, R. (2015). Fast rcnn, *IEEE Conference on Computer Vision and Pattern Recognition (CVPR)*, pp. 13–16.

Gundogdu, E. and Alatan, A. A. (2018). Good features to correlate for visual tracking, *IEEE Trans. Image Process.* **27**(5): 2526–2540.

He, Z., Fan, Y., Zhuang, J., Dong, Y. and Bai, H. (2017). Correlation filters with weighted convolution responses, *Proc. IEEE Conf. Comput. Vis. Pattern Recognit.* pp. 1992–2000.

Jia, Y., Shelhamer, E., Donahue, J., Karayev, S., Long, J., Girshick, R., Guadarrama, S. and Darrell, T. (2014). Caffe: Convolutional architecture for fast feature embedding, *Proceedings of the 22nd ACM international conference on Multimedia*, ACM, pp. 675–678.

Jurrasic Intro Dataset (2018). https://www.youtube.com/watch?v=lc0Ue hYemQA.

Kristan, M., Matas, J., Leonardis, A., Vojir, T., Pflugfelder, R., Fernandez, G., Nebehay, G., Porikli, F. and Čehovin, L. (2016).

A novel performance evaluation methodology for single-target trackers, *IEEE Trans. PAMI* **38**(11): 2137–2155.

Krizhevsky, A., Sutskever, I. and Hinton, G. E. (2012). Imagenet classification with deep convolutional neural networks, *Adv. Neural inf. Process. Syst.* **25**(2): 1097–1105.

LeCun, Y., Bengio, Y. and Hinton, G. (2015). Deep learning, *Nature* **521**(7553): 436.

Lin, T.-Y., Maire, M., Belongie, S., Hays, J., Perona, P., Ramanan, D., Dollár, P. and Zitnick, C. L. (2014). Microsoft coco: Common objects in context, *European conference on computer vision*, Springer, pp. 740–755.

Liu, W., Anguelov, D., Erhan, D., Szegedy, C., Reed, S., Fu, C.-Y. and Berg, A. C. (2016). Ssd: Single shot multibox detector, *European conference on computer vision*, Springer, pp. 21–37.

Oron, S., Bar-Hillel, A., Levi, D. and Avidan, S. (2015). Locally orderless tracking, *IJCV* **111**(2): 213–228.

Pal, S. K. (2018). Data science and technology: Challenges, opportunities and national relevance, *14th Annual Convocation Address, National Institute of Technology, Calicut, India, Sept. 29* .

Pal, S. K., Bhoumik, D. and Chakraborty, D. B. (2020). Granulated deep learning and z-numbers in motion detection and object recognition, *Neural Computing & Application*, **32**: 16533–16548.

Pal, S. K. and Chakraborty, D. B. (2017). Granular flow graph, adaptive rough rule generation and tracking, *IEEE Trans. on Cyberns.* **47**(12): 4096–4107.

Pal, S. K., Ghosh, A. and Shankar, B. U. (2000). Segmentation of remotely sensed images with fuzzy thresholding, and quantitative evaluation, *Int. J. Remote Sens.* **21**(11): 2269–2300.

Pal, S. K., Shankar, B. U. and Mitra, P. (2005). Granular computing, rough entropy and object extraction, *Pattern Recogn. Lett.* **26**(16): 2509–2517.

Patacchiola, M. and Cangelosi, A. (2017). Head pose estimation in the wild using convolutional neural networks and adaptive gradient methods.

Possegger, H., Sternig, S., Mauthner, T., Roth, P. M. and Bischof, H. (2013). Robust real-time tracking of multiple objects by volumetric mass densities, *IEEE Proc. on CVPR*.

Pramanik, A., Pal, S., Maiti, J. and Mitra, P. (n.d.). Granulated rcnn and multi-class deep sort for multi-object detection and tracking, *IEEE Trans. on Emerging Topics in Computational Intelligence* (*Accepted*).

Redmon, J., Divvala, S., Girshick, R. and Farhadi, A. (2016). You only look once: Unified, real-time object detection, *Proceedings of the IEEE conference on computer vision and pattern recognition*, pp. 779–788.

Ren, S., He, K., Girshick, R. and Sun, J. (2015). Faster r-cnn: Towards real-time object detection with region proposal networks, *Adv. Neural inf. Process. Syst.* pp. 91–99.

Visual Tracker Benchmark Data (2018). http://cvlab.hanyang.ac.kr/track er_benchmark/datasets.html.

Zadeh, L. A. (1996). Fuzzy logic= computing with words, *IEEE Trans. on fuzzy systems* **4**(2): 103–111.

Zadeh, L. A. (2011). A note on z-numbers, *Inf. Sci.* **181**(14): 2923–2932.

Zhong, W., Lu, H. and Yang, M.-H. (2014). Robust object tracking via sparse collaborative appearance model, *IEEE Trans. Image Process.* **23**(5): 2356–2368.

Chapter 7

Video Conceptualization

7.1 Introduction

We have so far addressed different challenges of video processing, like object tracking from videos, reliability measures of different tracking methods, and moving object recognition from video scenes. Here in the present chapter we are going to focus on a very important issue of video processing. It is event prediction (precognition) or recognition from videos. Along with some other recent methods we will discuss an unsupervised method for event prediction, namely, conceptualization (Chakraborty and Pal, 2021) in detail.

Video conceptualization, in other words, is a technique of pre-recognition of events in videos with object motion analysis. That is, determining whether there is a probability of some events taking place and further analysis of the content of the video is required or not can be decided with video conceptualization. This process is much simpler, requires no labeled data and is faster than video understanding techniques. Therefore, labeled data dependency and time consumption of an automated computer vision system (e.g., any surveillance system) can be reduced with this technique. This method is designed in a rough set theoretic granular computing framework where granules are formed over video sequences in two layers. The first layer of forming granules contains the spatio-color information of each frame. The second layer contains the motion information of those spatio-color granules. That is why these are named as 'motion granules'. The object–background rough set was defined over these motion granules. Accordingly, the moving object and background sets in the 'conceptualization' method were formed over these motion granules. These granules inherit the overlapping property, which helps to decrease the uncertainties in decision-making by pushing uncertain granules towards the boundaries of the sets. An uncertainty measure, namely, motion entropy is introduced in the 'conceptualization' method to quantify the uncertainty present in the motion of an object. The entropy values can reflect well whether there is any sudden change in the motion of the object, and can differentiate between continuous and random movements. The effectiveness of the concept of 'conceptualization' was tested with real-time video acquisition in an IoT set-up built with a raspberry pi 3B+ in lab environment.

The rest of the chapter is organized as follows. A few benchmark studies on video understanding and event detection are discussed in Section 7.2. The 'conceptualization' method is explained in details in Section 7.3 along with an IoT architecture for its realization/demonstration. This includes the formulation of two-layered motion granules, definition of object–background as rough sets over this granulation and the definition of motion entropy. The key steps and algorithms for video conceptualization are described in Section 7.4. Experimental results along with those in IoT setup are explained in Section 7.5 to demonstrate the effectiveness of the 'conceptualization' method. The overall conclusion is drawn in Section 7.6.

7.2   Related Work

There exist a significant amount of work that address the task of video understanding (Borges *et al.*, 2013). This problem was first dealt by Brand (1996) where videos of manipulation tasks were interpreted with psychological-based causal constraints to detect meaningful changes in motions. An approach of activity pattern analysis was then developed by Stauffer *et al.* (2000) with accumulation of information from multiple cameras. A method of automatic discovery of the key patterns of motion was formulated by Yang *et al.* (2009) by using a low level feature, pixel-wise optical flow, several of which were embedded later in a diffusion map framework. Behavior analysis with video understanding by tracking people and analyzing the trajectory with Mean-shift algorithm was formulated by Zaidenberg *et al.* (2013). An unsupervised method for video understanding was proposed by Milbich *et al.* (2017) where combinatorial sequence matching is performed to train a CNN (Convolutional

Neural Network), and thereby using a huge amount of labeled data for training the CNN. Mademlis *et al.* (2018) came up with a method of unsupervised video summarization with salient features. Another way of video understanding with desktop action recognition from ego-centric videos was developed by Cai *et al.* (2018). It was mainly focused on hand motion analysis. Another problem related to video understanding is re-identification of person in surveillance system. It was recently addressed by Gao *et al.* (2020) with pose-guided spatio-temporal alignment.

Event recognition from videos is a step ahead of inference of video understanding. There are several methods addressing this task. We are going to discuss here a few popular ones. Ke *et al.* (2007) developed a method of event understanding from a crowd by matching spatio-temporal segments among consecutive frames. The abnormal event detection and generation of description in a human-understandable format with a hybridized CNN model was developed by Himani *et al.* (2017). Rare event detection from satellite images with fine-tuned representation learning was developed by Hamaguchi *et al.* (2019). Human behavior recognition from multi-view camera was described by Hsueh *et al.* (2020) with a deep network that was developed by combining CNN and long-short-term memory network. Pre-recognition or prediction of future events, on the other hand, is a more challenging task that has been addressed recently. Liang *et al.* (2019) developed a technique of future event prediction by combining person-behavior and person-interaction together.

All of the aforementioned approaches either need initial manual intervention or a huge amount of labeled dataset for training. Therefore, the methods discussed above require different training data to be applied for different types of videos. But the 'conceptualization' method (Chakraborty and Pal, 2021), that we are going to discuss here to indicate the possibilities of events occurring and to identify the frames with maximum information, both with offline sequences and real-time sequences in IoT, does not need any.

7.3 Video Conceptualization and IoT Architecture

Here an unsupervised *rough set-based approach* is described that is able to conceptualize a video based on the nature of moving–static elements present in it and thereby infer the possibility of some future event taking place in the scene. The moving object(s) and background are represented as rough sets over a temporal domain. The granules are formed here in two layers. In the first layer, the spatio-temporal granules are formed as described in Section 1.7. In the second layer, the motion granules are formed over the spatio-temporal granules in the temporal domain. The object and background sets are approximated as rough sets over these motion granules. The decision about the continuous moving, random moving and sudden change (possible event) in a sequence are taken based on the nature of the neighborhood granules with respect to the sets. The nature of movements of the objects is then quantified with the newly defined motion entropy (M_E). The regions having high M_E values are detected as the regions with more information (where some unpredictable change occurs).

The basic operational principles of this conceptualization method are shown in Figure 7.1. Two layered motion granules are formed over the input video sequence in the first block (B-1). Rough set is defined over this granulation in the second block (B-2). Motion entropy is then measured for the granules present in the set in the third block (B-3). The output of this method is the video sequence with the moving object(s) classified according to the nature of their movements. The working principle of B-1 is described in Section 7.3.1. Operations of Blocks B-2 and B-3 are described in Sections 7.3.2 and 7.3.3, respectively. Here it is assumed that if there is a possibility of some event taking place there, then there would be some abrupt change in motion in a video scene. Therefore, the frames identified with sudden change may be labeled as the cause of possible future events.

IoT Architecture: The concept of conceptualization is tested with real-time video acquisition in IoT architecture. This architecture is shown in Fig. 7.2. It can be seen from Fig. 7.2 that real-time sequences are acquired with raspberry pi cam, and they are given as input to the raspberry pi 3B+ motherboard. The sequence is then analyzed with the conceptualization algorithm which is stored in raspberry pi. A breadboard with an LED is connected at the output of raspberry pi. The LED glows if any unusual motion is detected or possibility of some event occurring arises in the input sequence. Details of the experimentations performed with this architectural set-up to demonstrate the significance of 'conceptualization' are provided in Section 7.5.

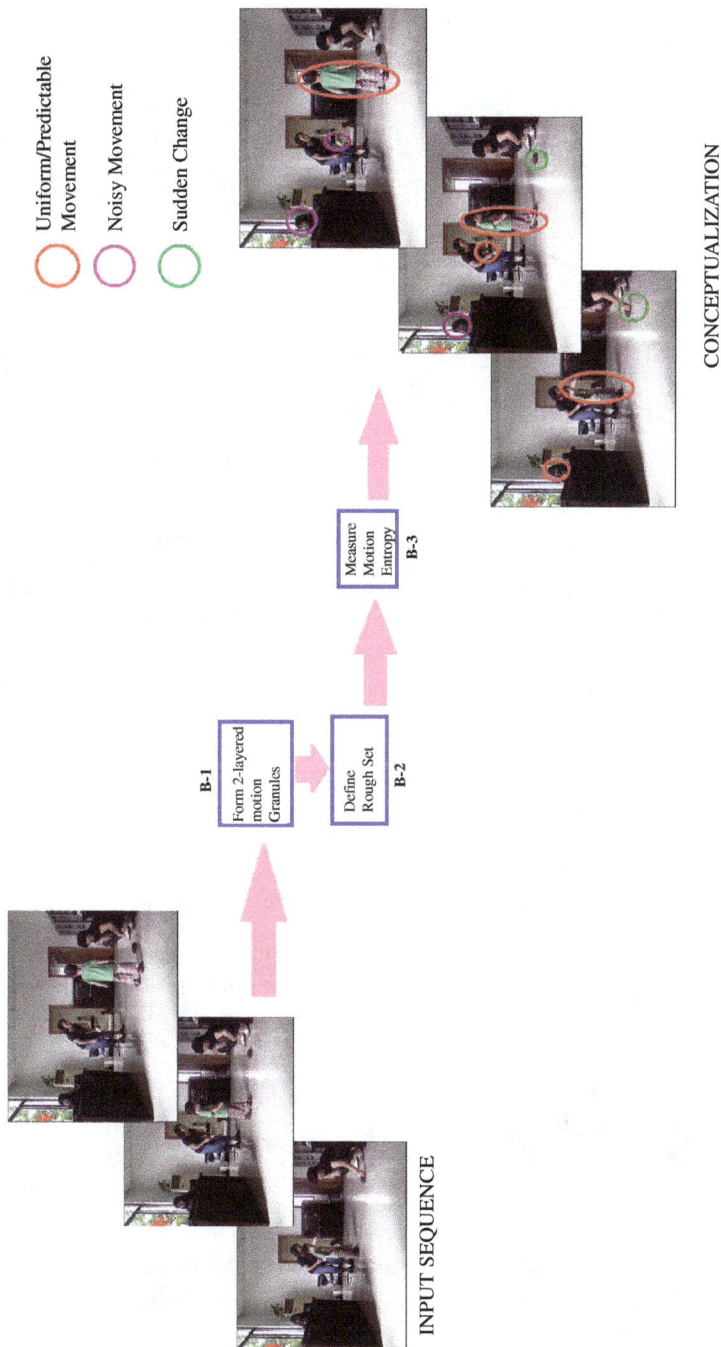

Fig. 7.1: Key steps of video conceptualization.

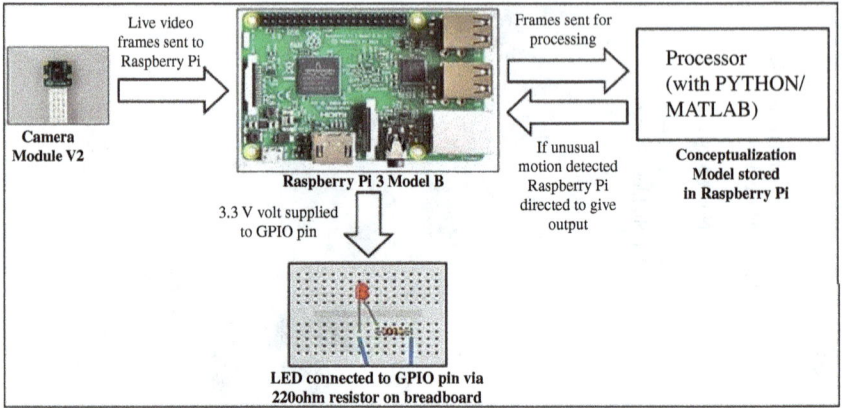

Fig. 7.2: IoT architecture for demonstrating video conceptualization.

7.3.1 *Formation of Two-layered Motion Granules*

Moving object(s) and background sets (upper and lower approximations) are defined (Chakraborty and Pal, 2021) using Pawlak's rough set (PaRS) (Pawlak, 1992) over neighborhood granules. These granules are constructed over temporal difference values of video frames. Spatio-temporal granules are formed in the first layer, and motion granules are formed in the second layer over these spatio-temporal granules. These layers are elaborated in the following sections.

7.3.1.1 *First Layer: Spatio-temporal Granules*

The formation of spatio-temporal granules is described in detail in Chapter 3. Here we describe it in brief for the convenience of readers. Let the current frame of a sequence be denoted as f_t (size $M \times N$) and its previous P-frames be $f_{t-p} : p = 1, \ldots, P$. The difference values are stored in P number of different matrices (T_V_p). The differences between f_t and all of its previous $1, \ldots, P$ frames are computed. It is shown in Eq. (7.1). T_V is of size $M \times N$.

$$T\_V_p = |f_t - f_{t-p}| \quad : \quad p = 1, \ldots, P. \tag{7.1}$$

The spatio-temporal granules are formed considering values of the points as T_V_p over the spatial domain. Let x_i be the position of a pixel in the difference frame (T_V_p), then the granule around it at p^{th} frame, denoted as $\aleph_{sp-tmp_p}(x_i)$, is formed according to Eq. (7.2).

$$\aleph_{sp-tmp_p}(x_i) = \bigcup x_j \in T\_V_p, \tag{7.2}$$

where x_i and x_j are binary connected over $|T\_V_p(x_i) - T\_V_p(x_j)| < Thr_t$ and $x_j \in T\_V_p$.

7.3.1.2 *Second Layer: Motion Granules*

The motion granules are defined to unify the similar spatio-temporal granules present in each difference frame T_V_p. That is, P number of $\aleph_{sp-tmp}(x_{ip})$ can be present there in each motion granule. Here x_{ip} is the location of the representative point of the spatio-temporal granule at frame T_V_p. The motion granules will contain the optical flow information of each $\aleph_{sp-tmp}(x_{i1})$.

One may note that $\aleph_{sp-tmp}(x_{ip})$ and $\aleph_{sp-tmp}(x_{i(p-1)})$ represent two regions in two different frames T_V_p and T_V_{p-1}, respectively. Now, if these two regions are plotted in the same frame, those regions may have non-zero intersection, or a region may be a subset to another region. This phenomenon was considered while defining the motion granules. A motion granule $\mathcal{M}(x_i)$ over the P frames in the sequence is defined in Eq. (7.3).

An example of forming motion granules over a sequence as defined in Eq. (7.3) is shown in Fig. 7.3. Here the granules formed over moving objects are only shown for the sake of simplicity. In the first part of this figure (in the left most side) the frames in D-feature space are shown. Absolute difference is taken from frame f_t to frames f_{t-1}, f_{t-2} and f_{t-3}. Three difference frames are obtained in this way and spatio-temporal granules are formed on the three difference frames. The three binary images $(T\_V_1, .., T\_V_3)$ in the rightmost side of Fig. 7.3 show these spaio-temporal granules. The motion granules are formed over these spatio-temporal granules as shown in the lower part of Fig. 7.3. Here two motion granules (represented with red and green dotted bounding box) are formed as two moving objects are present. The red dots and green dots are the representative points in each frame with which the motion granules are formed. From this figure, one can estimate how motion granules are formed taking into

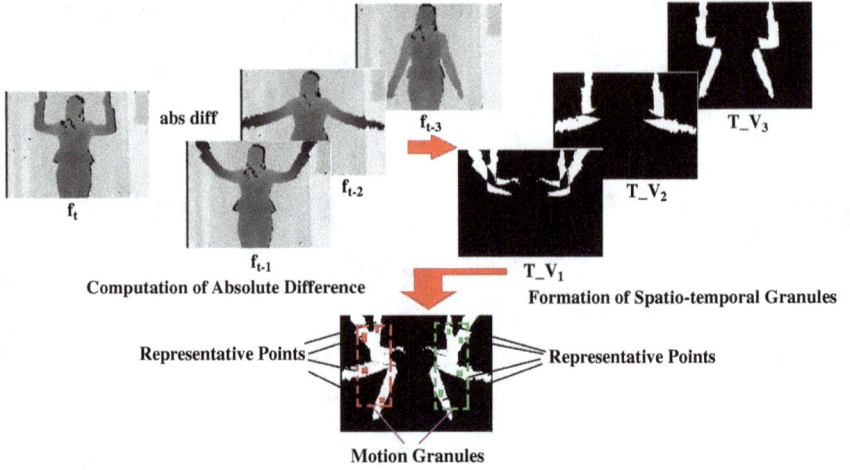

Fig. 7.3:   Example formation of motion granules.

account the optical flow of the spatio-temporal granules.

$$\mathcal{M}(x_i) = \{\bigcup \aleph_{sp-tmp_p}(x_{ip})\} \quad if \quad \aleph_{sp-tmp}(x_{i1}) \not\subset \aleph_{sp-tmp}(x_{ip}) \quad and$$
$$\aleph_{sp-tmp}(x_{i(p-1)}) \bigcap \aleph_{sp-tmp}(x_{ip}) \neq \emptyset \quad \forall p = 2, \dots, P.$$
$$(7.3)$$

Note that in Eq. (7.3) $\aleph_{sp-tmp}(x_{ip})$ denotes the spatio-temporal granules formed in the pth frame. For example, the white segments in the frames T_V_1, T_V_2 and T_V_3 in Fig. 7.3 denote $\aleph_{sp-tmp_1}(x_{i1})$, $\aleph_{sp-tmp_2}(x_{i2})$ and $\aleph_{sp-tmp_3}(x_{i3})$ for the example frames. Two motion granules, formed over these spatio-temporal granules according to Eq. (7.3), are shown with red and green dotted rectangles in Fig. 7.3.

7.3.2 *Object–Background as Rough Sets*

Here the moving object(s) and background are defined as a rough set over the aforesaid motion granules in the current frame f_t. The information from the set of its previous P number of frames are taken into account and represented as $\{P\} = \{t-1, \dots, t-P\}$. The feature of a granule $\mathcal{M}(x_i)$ that is considered here to define the sets is the signed Manhattan Distance (along x and y axes) between the consecutive points of $\mathcal{M}(x_i)$. This distance is denoted as $\vec{d_p}$. One may note that $\vec{d_p}$ is a vector with four components, viz, two absolute

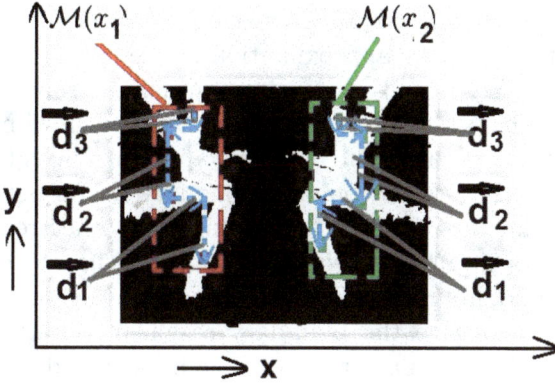

Fig. 7.4: A pictorial example of distance computation following Eq. 7.4 over two motion granules $\mathcal{M}(x_1)$ and $\mathcal{M}(x_2)$.

values (along x and y axes) and two signs (positive or negative in those axes). It is computed according to Eq. (7.4)

$$\overrightarrow{d_p} = x_{i(p-1)} - x_{ip}, \forall p = 2, \ldots, P. \tag{7.4}$$

Figure 7.4 represents the computation of $\overrightarrow{d_p}$s over two motion granules $\mathcal{M}(x_1)$ and $\mathcal{M}(x_2)$ pictorially. Here the dotted blue lines with directions signify the distances between the consecutive points in a granule. That is, the signed distance along both the x and y axes are computed here. One may note that all the d_1, d_2 and d_3 contain two magnitudes and two directions (positive or negative) in both the granules.

A granule $\mathcal{M}(x_i)$ is said to belong to lower approximation of the object region, if the absolute values of all $|d_p|$s are greater than zero and $sign(d_1) = sign(d_2) = \cdots = sign(d_P)$. That is, the granules with continuous motion (in the continuous moving object(s)) will belong to $\underline{O_t}$ (see Eq. (7.5(a))). If there exist at least one pair of points for which $|d_p| > 0$ in a granule ($\mathcal{M}(x_i)$) and $sign(d_p)$s are the same at least twice over $p \in \{P\}$, then the granule will belong to the upper approximation of the object region in f_t $\overline{O_t}$ (see Eq. (7.5(b))). That is, any object in motion (may or may not be continuous) will belong to the upper approximation of the object set. Similarly, while forming the background set, $|d_p| = 0$ for all $p \in \{P\}$ for a granule $\mathcal{M}(x_i) \in \underline{B_t}$, where $\underline{B_t}$ stands for the lower approximation of the background region in the frame f_t (see Eq. (7.5(c))). If there exists at least one point in a granule for which $|d_p| = 0$, then the granule will belong to the upper approximation of the background region in f_t, i.e., $\overline{B}_t$ (see Eq. (7.5(d))). A crisp and definite separation can be performed in this way in the lower approximated regions. However, there exists overlapping in the boundary regions of both the sets. One may note that there is no common boundary region between object and background sets. Rather, there exist two other regions or two different sets of granules solely characterizing the object boundary and background boundary. That means, there are a few granules which solely belong to the boundary region of the object, but do not belong to the boundary region of the background. The reverse is also true for some other granules. Besides, neither of these granules belong to the lower approximations of the sets, as there exists a non-zero probability to belong to its complementary set. Here comes the concept of the 'other class', that is the class which is neither object nor background. The detection of this class can be treated as noise, i.e., no relevant information is present there in these regions even though some motion may be there. This phenomenon of the newly defined object–background set helps to understand the video more clearly.

$$\underline{O}_t = \{\mathcal{M}(x_i) \in U : \ |d_p| > 0 \quad \forall p \in \{P\} \ \&$$

$$sign(d_1) = sign(d_2) = \cdots = sign(d_P)\} \qquad (7.5a)$$

$$\overline{O}_t = \{\mathcal{M}(x_i) \in U : \ |d_p| > 0 \quad \exists p \in \{P\} \ \&$$

$$sign(d_{pm}) = sign(d_{pn}) \ while \ \{pm, pn\} \in \{P\}\} \qquad (7.5b)$$

$$\underline{B}_t = \{\mathcal{M}(x_i) \in U : \ |d_p| = 0 \quad \forall p \in \{P\}\} \qquad (7.5c)$$

$$\overline{B}_t = \{\mathcal{M}(x_i) \in U : \ |d_p| = 0 \quad \exists p \in \{P\}\}. \qquad (7.5d)$$

A pictorial representation of the rough set (defined in Equation (7.5)) is shown in Fig. 7.5 in a two-dimensional feature space. Here a two-class based rough set is shown with overlapping granules. The '+'-class represents the object class, whereas the '−'-class represents the background class. How the object boundary and background boundary regions overlap each other and yet not be the same is visualized here . Algorithm 5 describes the steps for formation of rough set over video.

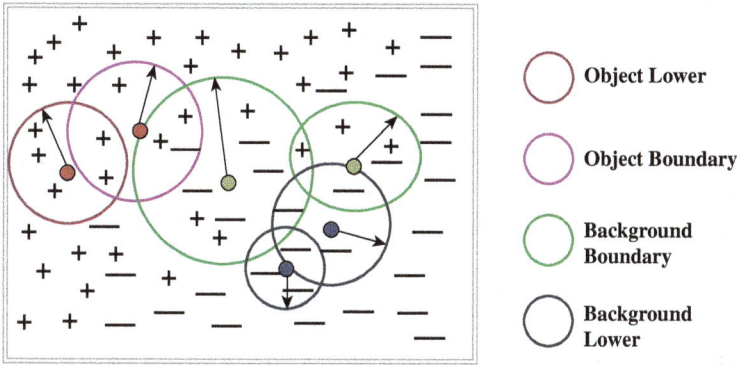

Fig. 7.5: Pictorial representation of a two-class (object–background) rough set.

7.3.3 *Motion Entropy*

This entropy measures the amount of certainty in the movements of the object(s). One may note that another motion uncertainty measure, namely, intuitionistic entropy, is also described in this book in Chapter 4. But intuitionistic entropy mainly determines the certainty of a moving object to be present in a location at a given time. Whereas the motion entropy, to be described here, determines the uncertainties present in the motion pattern of an object itself. Here it is assumed that if there is some probability of an event taking place, the motion pattern of moving/static object would face sudden change and thereby cause an uncertainty. Motion entropy concerns computation of that uncertainty. It consists of two uncertainty measures, namely, velocity entropy and acceleration entropy. As shown in Algorithm 6, the decision-making regarding the movements is performed based on the nature of shifts from frame-to-frame. Both the static object(s) and object(s) with continuous movements are supposed to have low motion entropy values. Whereas the regions with sudden change (static-to-moving or moving-to-static) would have higher entropy values. The information regarding those regions with uncertainty in the system should be updated after observing their nature in the next P frames. One may note that, this entropy is mainly computed over the uncertain regions where motion is or may be present, i.e., the regions present in $\underline{O}_t \bigcup \{\overline{O}_t - \underline{O}_t\} \bigcup \{\overline{B}_t - \underline{B}_t\}$. *Shift* and *SC* (Eq. (7.6)) are considered during the formulation of this measure.

Algorithm 5 Formation of Rough Set over Video

INPUT: $f_t, \ldots, f_{t-P}, Thr_t$
OUTPUT: $\underline{O}_t, \overline{O}_t, \underline{B}_t, \overline{B}_t$
INITIALIZE: $O_t \Leftarrow \emptyset\ B_t \Leftarrow \emptyset$
1: Form first layer granule, i.e., spatio-temporal granules $\aleph_{sp-Tmp}(x_i)$ according to Eq. (7.2).
2: Form second layer granule, i.e., motion granule $\mathcal{M}(x_i)$ according to Eq. (7.3).
3: Compute the number of points covered by each $\mathcal{M}(x_i)$ and sort these according to size (large to small).
4: Find the granule $\mathcal{M}_L(x_i)$ with maximum no. of points $x_i \in \mathcal{M}_L(x_i)$.
5: Compute $\vec{d_p}$ for $\mathcal{M}_L(x_i)$ according to Eq. (7.4).
6: Put it to the any one of the sets $\underline{O}_t, \overline{O}_t, \underline{B}_t, \overline{B}_t$ according to Eq. (7.5).
7: Find the next $\mathcal{M}, (\mathcal{M}_{L-1})$. Set L − 1=L.
if $\mathcal{M}_{L-1} \subset \mathcal{M}_L$ **then**
    Remove $\mathcal{M}_{L-1}$ form the set and go to step 5.
else
    Set L − 1 = L and go to step 4.
end if
8: Do it for all the $\mathcal{M}$.
9: Detect the no. of separate moving objects by computing the spatial nearness among $\mathcal{M}(x)$s in $\underline{O}_t, \overline{O}_t$ and $\{\overline{B}_t - \underline{B}_t\}$.

Let M represent a moving object in the f_t, i.e., $M_t \in \overline{O}_t$ or $M_t \in \overline{B}_t - \underline{B}_t$. Its nature of movement (amount of shifts and changes in shifts) over P frames are stored in two matrices, namely, $Shift(M)$ and $SC(M)$. The matrix $Shift(M)$ contains the location change information of moving object M in consecutive frames. The matrix $SC(M)$ stores the change in $Shift(M)$ over time. In other words, $Shift(M)$ contains the velocity information and $SC(M)$ contains the acceleration information of the object M. These matrices are defined in Eq. (7.6).

$$Shift(M) = \{Shift_p : p = 1, \ldots, P\} : Shift_p = M_{t-p} - M_{t-(p-1)}$$
(7.6a)

$$SC(M) = \{SC_p : p = 1, \ldots, P\} : SC_p = \frac{d}{dt}(Shift(M)), \quad (7.6b)$$

Algorithm 6 Moving Pattern Recognition of mth Object Till t^{th} Frame and Event Precognition

---

INPUT: $M_t, \ldots, M_{t-P}$
OUTPUT: Decision regarding the movement pattern
INITIALIZE: $D \Leftarrow \emptyset$
1: Compute spatial shifts of $Moving$ regions.

$$Shift_p = M_{t-p} - M_{t-(p-1)}$$

to form the set $Shift = \{Shift_k : k = 1, \ldots, P\}$.
2: Compute the change in $Shift$ values by computing its derivative over time:

$$SC_p = \frac{d}{dt}(Shift)$$

and where the set $SC = \{SC_k : k = 1, \ldots, P\}$.
3: Compute ME for M_t.
if $M_t \in \underline{O}_t$ or if $M_E \leq 0.2$ **then**
    $D = ContMov$
else if $M_t \in \overline{O}_t$ and $0.2 < M_E \leq 0.9$ **then**
    $D = RndMov$
else if $M_t \in \overline{B}_t$ and $0.9 < M_E \leq 1$ **then**
    $D = SuddChng$
end if
4: Label M_t to cause a probable event if $D = SuddChng$.

---

where M_{t-p} denotes the location of the moving object M in the frame f_{t-p}. One may note that, $Shift_p$ is also a signed *Manhattan Distance* between objects in consecutive frames and computed similarly as of $\overrightarrow{d_p}$ in Eq. (7.4).

Let the mean values of the matrices be represented as $Shift_m$ and SC_m. Let $ShiftL$ and SCL be two matrices such that, $ShiftL = \{s : s \in Shift(M) \& s \geq Shift_m\}$ and $SCL = \{sc : sc \in SC(M) \& sc \geq SC_m\}$.

The velocity roughness (V_R) and acceleration roughness (A_R) of that region is computed as:

$$V_R = 1 - \frac{|ShiftL|}{|Shift(M)|} \tag{7.7a}$$

$$A_R = 1 - \frac{|SCL|}{|SC(M)|}, \qquad (7.7b)$$

where $|.|$ represents the cardinality of the set. Motion roughness is thereby defined as:

$$M_R = \frac{V_R + A_R}{2} \qquad (7.8)$$

and the motion entropy (M_E) is:

$$M_E = M_R \times e^{1-M_R}. \qquad (7.9)$$

The variations in the values of the uncertainty measures over different types of movements are shown in Table 7.1 with three examples: continuous movement/with uniform change in motion (i.e., predictable), random movement and sudden change in movement.

The moving patterns shown in Table 7.1 are ideal by nature. In real life, this is expected to be more complex. However, one can have some ideas regarding the nature of the movements by observing the values of M_R and M_E. Definite differentiation among the noise and moving object(s) and sudden changes is possible after observing the corresponding M_E.

One may note that the values of M_E are expected to be the highest ($\simeq 1$) in case of sudden change. It shows that maximum information will be present there due to its unpredictability. Therefore, while detecting the possible event(s), the threshold of M_E was set close to 1, i.e., 0.9. On the other hand, $M_E \simeq 0$ when the movement is predictable, that is, very less information is available there. Therefore, the threshold of M_E-value was set close to 0, at 0.2, while detecting the predictable motion.

7.4 Conceptualizing the Video by Estimating the Moving Patterns of Objects and Precognized Events

The basic steps behind the 'conceptualization' method are shown in Fig. 7.6. This method mainly consists of two parts. The first one is unsupervised formation of object–background sets and the second one is categorizing the moving segments by observing the nature of the movement of the object(s). The formation of RS (rough set) is described in Algorithm 5.

Table 7.1: Variations in motion entropy with different movement patterns.

Movements	*Shift* Pattern	*SC* Pattern	M_R	M_E
Predictable $\in \underline{O}_t$			0.53	$\simeq 0$
Sudden Change			0.95	0.998
Random $\in \overline{O}_t$			0.5	0.82

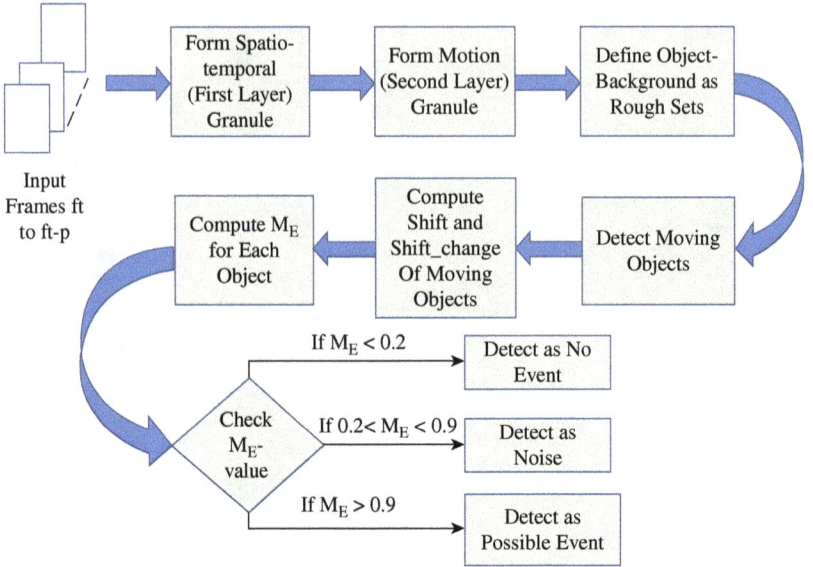

Fig. 7.6: Step-wise flow diagram of video conceptualization.

After the formation of the object–background sets, the objective is to estimate the nature of motion and change in that nature with respect to time in each set. This is done by mapping and analyzing the frame-to-frame shifts of each object(s) belonging to the sets $\underline{O}_t$, $\{\overline{O}_t - \underline{O}_t\}$ and $\{\overline{B}_t - \underline{B}_t\}$. No operation will be conducted for the lower background regions, until and unless there occurs some frame-to-frame deviation in that region. The pattern of movement of certain objects is primarily estimated from the initial P frames and then the regions $\underline{O}_t$ and $\overline{O}_t$ get updated alongwith the stream. Let the Mth moving region in f_t be denoted by M_t ($M_t \in \overline{O}_t \cup \{\overline{B}_t - \underline{B}_t\}$). Let the decision attributes be $D = \{ContMov, RndMov, SuddChng\}$ representing continuous moving ($ContMov$), random moving ($RndMov$) and suddenly changed moving ($SuddChng$) objects. It was assumed (Chakraborty and Pal, 2021) that if any sudden change of motion takes place in an object, then there is a high probability of some event occurring. The movement–pattern estimation and event precognition are done according to Algorithm 6.

In Algorithm 6, $Shift_p$s are signed integers along X and Y axes, and therefore it is also a vector like d_p (in Eq. (7.4)) with two elements. The value represents the shift and the sign represents the direction. Therefore, if the sign of the values in the set *Shift* remains almost constant, then the deviation in the set *SC* is very low which reflects continuous motion of the object. Whereas, if there is some probability of an event taking place, then a sudden change in object motion in f_t will be observed. It will cause both of the sets *Shift* and *SC* to have only one non-zero value and those are of same magnitude. The performance of this algorithm is demonstrated with experimental results in the next section.

7.5 Results and Discussions

7.5.1 *Preliminary Assumptions and Parameter*

The results on video conceptualization presented here are presented under the assumption that all the videos are captured by a static camera and there is no occlusion or overlapping. Selection of parameters is not a very crucial issue here. One of the objectives of the study is to make the parameters adaptive, as much as possible, so that these can take the approximate values automatically depending on the nature of the movement and size of the objects in a sequence. For example, the value of P (the number of previous frames to be considered) was chosen depending upon the speed of the object. Let τ_p be the binarized T_V_p (see Eq. (7.1)) over a threshold Th. Then, the lowest value for which $\bigcap \tau_p : p = 1, \ldots, P \equiv \bigcap \tau_p : p = 1, \ldots, P+1$ is chosen as P. The threshold value Thr_t used in Eq. (7.2) was set to $0.3 \times Median(T\_V_p)$, where $Median$ stands for the statistical median.

7.5.2 *Experimental Results in PC*

The 'conceptualization' method is primarily implemented in MATLAB on a PC with 3.4-GHz CPU. The main advantages of this algorithm are as follows:

(i) unsupervised conceptualization of videos by analyzing their nature of the movements, and

(ii) detecting the sudden changes or regions with more information with motion entropy.

The video sequences with different characteristics, e.g., indoor/outdoor surveillance (Davis and Sharma, 2007; PETS-2001, 2001; AVSS-2007, 2007; Possegger *et al.*, 2013), single/multiple moving object(s) (Song and Xiao, 2013; Davis and Sharma,

2007; Possegger *et al.*, 2013), and body part(s) movements (ChaLearn, 2011) were considered during the experimentation. All of the video images used are in RGB color space, however D-feature sensed by kinect sensor (Han *et al.*, 2013) was also used wherever it was available (ChaLearn, 2011; Davis and Sharma, 2007; Song and Xiao, 2013). The algorithm was executed almost over 1000 frames in total, however only a few results are presented here, as examples.

The video sequences over which the results are shown here have the following characteristics. Sequence $P - 01$ (PETS-2001, 2001) initially contains one moving person and then different moving cars from different directions appear one by one with different velocities and with variation in shape and size over the sequence, one of them get stopped and become part of the background. There are two people moving in different directions, who stop, move only hands, again start moving in $5b$-sequence (Davis and Sharma, 2007). There is initially one person moving with a bag in a railway platform, where the background train disappears slowly and the person stops and starts to move his hands in $A - 07$ sequence (AVSS-2007, 2007). Sequence *Child* (Song and Xiao, 2013) contains one child walking continuously, whereas the people around him move partially or randomly. One lady is moving her hands continuously with a few movements in the rest of the body in M_4 (ChaLearn, 2011). Two people enter one by one and change their directions, then one of them suddenly sits on the floor and the other one jumps over him in the sequence $cam - 132$ (Possegger *et al.*, 2013). The results that are obtained with these datasets are shown in the following sections.

The visual results are represented as:

(i) Continuous movement/without any probable event: marked in red color,

(ii) Random movement/noise: marked in purple color,

(iii) Sudden change/probable event: marked in green color.

Four frames for each sequence are shown in Fig. 7.7 to demonstrate the process of decision-making. It can be seen that the same object(s) is labeled in different classes in different frames depending on how its movement varies. For example, the moving car in frame 449 (Fig. 7.7(1-b)) is labeled as sudden change (moving-to-static) in the frame 669 (1-c), and then to background 823 (1-d). Similar process was executed in the other sequences over objects with similar

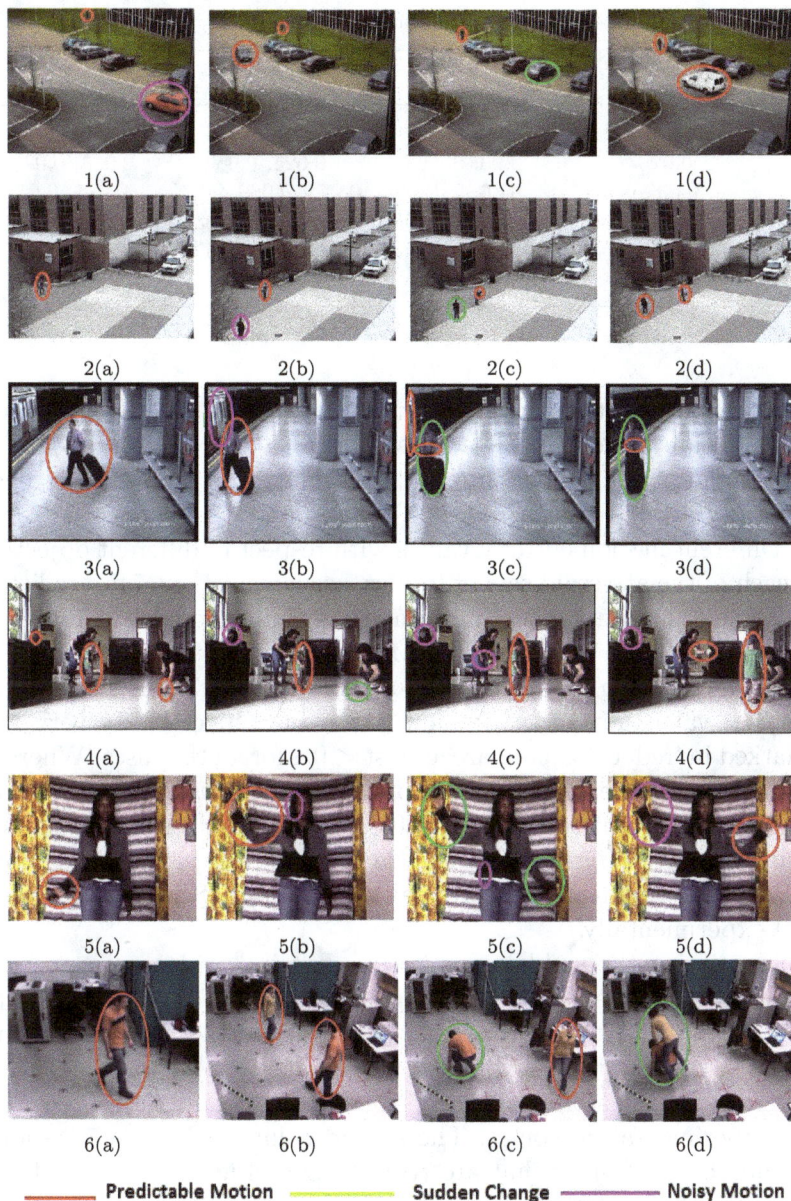

1(a) 1(b) 1(c) 1(d)

2(a) 2(b) 2(c) 2(d)

3(a) 3(b) 3(c) 3(d)

4(a) 4(b) 4(c) 4(d)

5(a) 5(b) 5(c) 5(d)

6(a) 6(b) 6(c) 6(d)

———— Predictable Motion ———— Sudden Change ———— Noisy Motion

Fig. 7.7: Conceptualization results for frame nos. (1) 12, 449, 669, 823 from $P-01$ sequence, (2) 196, 364, 746, 1418 from $5b$ sequence, (3) 2339, 2389, 2445, 2492 from $A-07$ sequence, (4) 5, 25, 44, 69 from $Child$-sequence, (5) 7, 16, 21, 28 from M_4 sequence and (6) 159, 239, 310, 365 form $cam-132$ sequence.

Table 7.2: M_E values for the frames of Fig. 7.7.

	(a)	(b)	(c)	(d)
1	0.81, 0.04	0.07, 0.02	0.94, 0.06	0.05, 0.02
2	0.02	0.03, 0.74	0.025, 0.96	0.034, 0.026
3	0.033	0.06, 0.81	0.044, 0.069, 0.97	0.075, 0.91
4	0.02, 0.058, 0.043	0.76, 0.07, 0.94	0.73, 0.67, 0.03	0.65, 0.08, 0.02
5	0.036	0.023, 0.78	0.93, 0.78, 0.99	0.82, 0.08
6	0.053	0.033, 0.12	0.97, 0.077	0.98

movement characteristics. The random movements/newly appeared object(s) are present in the scenes (1(a), 3(b), 4(b), 4(c), 5(b)–(d)). The moving-to-static changes/sudden changes are present in scenes (1(c), 2(c), 3(c)–(d), 4(b), 5(c), 6(c) and 6(d)).

Different motion entropy values with respect to different object(s) in each of the aforesaid sets of frames, as shown in Fig. 7.7, are listed in Table 7.2. The frames are named as those indexed in Fig. 7.7, (i.e., from 1(a) to 6(d)) and the M_E values for different regions of a frame are listed in the respective location of Table 7.2. It is seen that the M_E values for the continuously moving object(s) are very low (marked in 'red' color) and are almost < 0.1 for all the cases. Whereas the random movements/newly appeared object(s) have M_E values (marked in purple) between 0.6 to 0.85. The M_E values for the sudden change object(s)/probable events (marked in green) are always above 0.9. These validate the theoretical assumptions (Sections 7.3.3 and 7.4) experimentally.

Identifying the sudden changes present in a video sequence has always been crucial and poses a big challenge. In Table 7.3, the accuracy of the said task with 'conceptualization' method (Algorithm 6) is shown. The frames with sudden changes (FSD) were identified manually for the aforementioned sequences and validated those with the experimental outcomes. The second column of Table 7.3 shows the number of frames that are truly FSD and identified as FSD by Algorithm 6, i.e., true positive (TP) frames in other words. The third column of this table shows the number of frames that are not FSD but identified as FSD by the algorithm, i.e., false positive (FP). The fourth column shows the accuracy of the algorithm in identifying sudden change, i.e., $TP/(TP + FP)$.

Table 7.3: Accuracy of Algorithm 6 in identifying FSD.

Name of the Sequence	TP Frames	FP Frames	Accuracy (%)
$P - 01$	65	6	91
$5 - b$	22	4	84
$A - 07$	56	5	92
Child	11	2	85
M_4	65	12	84
$cam - 132$	78	8	90

Table 7.4: Percentage of number of frames reduced from further processing.

Name of the Sequence	Total number of frames	Total number of frames with $M_E > 0.9$	% of gain
$P - 01$	350	65	81.15
$5 - b$	130	22	85.33
$A - 07$	250	56	77.6
Child	85	11	87.65
M_4	65	12	81.5
$cam - 132$	460	78	83

One may note that, the 'conceptualization' method does not generate any false negative frames. That means, the frames with true FSD will always be detected by the said algorithm.

It is claimed that the 'conceptualization' technique will cause computational gain as less number of 'highlight' frames will be required for further processing. The proof is shown quantitatively in Table 7.4. It has been previously discussed that the FSD will have the highest entropy value and they carry the useful information. Accordingly, it is shown how many frames of the aforementioned sequences have motion entropy values > 0.9 in the total sequence. From Table 7.4 one can see that the useful information is present only in 15–25% frames of various types of sequences under consideration. Therefore, further processing of 85–75% of the frames is not required to interpret the storyline of the videos. This is how huge computational gain can be achieved in video understanding.

7.5.3 *Comparative Results*

The unsupervised video conceptualization, comprising object tracking followed by event precognition, is a new concept. There is almost no other similar study to compare its performance. However, tracking based on object–background separation is an underlying part of conceptualization. Therefore, the comparative studies shown here are based on the moving object–background separation, described in Algorithm 5. The study is performed with a few recent robust tracking methods over the aforesaid datasets. With these comparative studies one will be able to judge how efficient the 'conceptualization' method is for the task of event precognition, since tracking performance is crucial for this task. The tracking methods with which the comparative studies are performed include NRBFG (Pal and Chakraborty, 2017) and follow six popular methods.

DeepTrack (Li *et al.*, 2016): In this method, a single CNN was used for tracking. This network was developed for learning effective features for the target object representation in a purely online manner.

CNT (Zhang *et al.*, 2016): Here simple two-layer convolutional networks were used for object representations. A 'deep network' structure was then introduced in visual tracking.

KPSR (He *et al.*, 2017): Key patch sparse representation-based tracker was introduced here. Sparse representation, selection of the key patches and designing the contribution factor of the patches were the basic steps of tracking here.

CST (Wang *et al.*, 2018): A constrained graph labeling algorithm was proposed here for tracking. Superpixel based transductive learning, the appearance fitness constraint and the temporal smoothness constraint are incorporated in the graph labeling algorithm which is then used for tracking.

DNNT (Chi *et al.*, 2017): A dual deep network is designed here for tracking. Its objective is to exploit the hierarchical features in different layers of a deep model and design a dual structure to obtain features from various streams. This model is updated online based on the observation tracked object in consecutive frames.

SIMT (Kwon and Lee, 2014): This method obtains several samples of the states of the target and the trackers during the sampling process using Markov Chain Monte Carlo (MCMC) method. Here the trackers interactively communicate and exchange information with others; thereby improving its performance and increasing the overall tracking performance.

The time and accuracy comparisons among these tracking methods with respect to that of the 'conceptualization' method are shown in Table 7.5. The accuracy is measured based on the distance between the centroids (CD) of the ground truth and the obtained tracked region of the respective frames. The ground truths were available along with the datasets that were used in the experiment. The time given here is the average CPU time in seconds needed to process a frame.

Table 7.5 shows that the tracking part of the 'conceptualization' method results in computational gain as less computation time is required compared to the other methods. It can also be noted that the accuracy of this method is not always the best among the others. Therefore, the method could be useful where no prior knowledge is available and faster decision-making is required.

7.5.4 *Experimental Results in IoT*

The effectiveness of the 'conceptualization' method is demonstrated here in an IoT architecture. The architecture is built by embedding a raspberry pi 3B+ motherboard, a raspberry pi cam for real-time data acquisition, and a breadboard with LED. The codes to implement these algorithms and techniques are written here in Python programming language by leveraging the predefined functions present in OpenCV library, Scipy. The detail of this architecture is shown in Fig. 7.2 at Section 7.3. The videos used here for experimentation are both the offline data, as described in the previous section, and the real-time data. The real-time (on-line) sequences were shot by moving a pen with different speeds and directions in front of the raspberry pi cam. In Fig. 7.8, two example frames for off-line and on-line videos with unusual change detection are shown. LED glows by detecting unusual movements. In case of on-line data, it glows whenever some change is found in the motion/movement of the pen.

In Table 7.6, the performance of 'conceptualization' (Algorithm 6) is demonstrated quantitatively. In the first six rows, Table 7.6 is compared with Table 7.4 in terms of the reduction of frames. The count in the third column of Table 7.6 is taken as the total number

Table 7.5:　Time and accuracy comparisons.

Sequence	Metric	NRBFG	Deep Track	CNT	KPSR	CST	DNNT	SIMT	Conceptualization
P – 01	CD	4.78	3.22	3.56	3.43	3.82	3.81	3.93	4.95
P – 01	Time	0.325	0.305	0.320	0.332	0.381	0.307	0.368	0.165
A – 07	CD	4.54	5.94	5.86	5.21	4.68	4.76	4.81	5.21
A – 07	Time	0.234	0.323	0.366	0.383	0.265	0.271	0.231	0.189
Child	CD	4.23	5.32	4.22	3.57	3.82	3.72	3.61	4.51
Child	Time	0.308	0.351	0.385	0.335	0.381	0.344	0.276	0.198
M_4	CD	2.72	2.02	2.42	1.65	1.42	1.55	1.36	2.92
M_4	Time	0.256	0.241	0.295	0.222	0.251	0.246	0.271	0.111
cam – 132	CD	4.82	6.02	5.42	4.65	4.72	4.58	4.75	5.22
cam – 132	Time	0.276	0.421	0.415	0.322	0.351	0.316	0.284	0.192

Detection of Unusual Change in Video

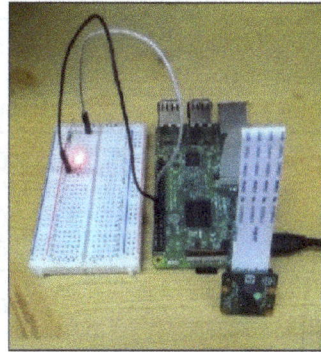

LED Glows

(a)

LED Glows by Detecting Unusual Change in Real-time Video

(b)

Fig. 7.8: LED glows in IoT by detection of unusual change in motion: (a) frame no. 692 of $P - 01$ sequence and (b) a frame in real-time sequence.

of times the LED glowed in each sequence, and it is compared with the total number of frames with $M_E > 0.9$ of Table 7.4 for the same sequence. It can be seen from Table 7.6 that the number of frames where LED glowed IoT is almost the same as that obtained in the PC. One to two frames only got lost in each sequence due to IoT set-up, but the performance is almost similar. Therefore, it can be concluded that the 'conceptualization' method performs well in IoT architecture. The last two rows of Table 7.6 show the performance of this IoT

Table 7.6:   Comparative study on frame number reduction.

Name of the Sequence	Total number of frames	Total number of frames where LED glows	Total number of frames with $M_E > 0.9$
$P - 01$	350	62	65
$5 - b$	130	20	22
$A - 07$	250	55	56
Child	85	10	11
M_4	65	12	12
$cam - 132$	460	76	78
$Real - TimeSq - 1$	90	46	48
$Real - TimeSq - 2$	110	32	33

architecture with two real-time sequences. The real-time sequences, as mentioned before, were shot by moving a pen with different speeds and directions in front of the raspberry pi cam. The LED glows when there is some change in motion of the pen. The same sequences were recorded and tested with M_E values in PC. The fourth column shows these results. From the last two rows of Table 7.6 the similar conclusion can be drawn for real-time sequences as that of the offline sequences. That is, the conceptualization algorithm is effective with real-time data acquisition, too, in IoT.

7.6   Conclusions and Discussions

In this chapter, a new video processing task, called 'video conceptualization', is described. The process involves rough set-based moving object–background classification and motion uncertainty analysis with motion entropy. The primary goals of this new application lie in predicting some event to occur, identifying the frames where some events may take place and selecting a few sets of consecutive frames as the highlight of the entire sequence. The effectiveness of the theories and methodologies described here is experimentally validated with different types of video sequences. The 'conceptualization' method is also applied and tested in an IoT framework where it proves to work properly even with real-time data acquisition. To our knowledge, no such unsupervised video conceptualization approach exists in literature. The application of event prediction in IoT is also very scarce. Therefore, no suitable comparative survey could be provided for the entire task, and comparative results are shown for the tracking part only. The unsupervised tracking method (Algorithm 5), which is a crucial part of the 'conceptualization' task, proves to be faster than several recent robust tracking methods. Besides, the motion uncertainty measure, viz. 'motion entropy', defined here can reflect which object(s) present in a video sequence have unpredictable movements, i.e., more suspicious behavior. Since the method is primarily developed for the videos captured with static cameras, it may not work if camera movement is present there in the video sequence.

This method could be extended to identification of object-of-events and description generation of possible events. Techniques to handle camera motion could also be integrated with this method to make it more robust. Further, the theories and methodologies described here may be applied in the other areas of computer vision, e.g., video summarization and shot boundary detection. Above all, this investigation could be highly effective in the IoT applications with video processing, like in designing smart city and smart home.

Bibliography

AVSS-2007 (2007). *Fourth IEEE Int. Conf. Adv. Video & Signal Based Surveillance.*

Borges, P., Conci, N. and Cavallaro, A. (2013). Video-based human behavior understanding: A survey, *IEEE Trans. on CSVT* **23**(11), 1993–2008.

Brand, M. (1996). Understanding manipulation in video, *Proceedings of the Second Intl Conf on AFGR*, IEEE, Killington, VT, pp. 94–99.

Cai, M., Lu, F. and Gao, Y. (2018). Desktop action recognition from first-person point-of-view, *IEEE Trans. on Cyberns.* 1–13. DOI: 10.1109/TCYB.2018.2806381.

Chakraborty, D. B. and Pal, S. K. (2021). Rough video conceptualization for real-time event precognition with motion entropy, *Information Sciences*, **543**, 488–503.

ChaLearn (2011). *ChaLearn Gesture Dataset (CGD 2011)*, California.

Chi, Z., Li, H., Lu, H. and Yang, M.-H. (2017). Dual deep network for visual tracking, *IEEE Tran. on Image Proc.* **26**(4), 2005–2015.

Davis, J. and Sharma, V. (2007). Background-subtraction using contour-based fusion of thermal and visible imagery, *Computer Vision and Image Understanding* **106**, 162–182.

Gao, C., Chen, Y., Yu, J.-G. and Sang, N. (2020). Pose-guided spatiotemporal alignment for video-based person re-identification, *Information Sciences* **527**, 176–190.

Hamaguchi, R., Sakurada, K. and Nakamura, R. (2019). Rare event detection using disentangled representation learning, *The IEEE Conference on Computer Vision and Pattern Recognition (CVPR)*.

Han, J., Shao, L., Xu, D. and Shotton, J. (2013). Enhanced computer vision with microsoft kinect sensor: A review, *IEEE Trans. on Cyberns.* **43**(5), 1318–1334.

He, Z., Yi, S., Cheung, Y.-M., You, X. and Tang, Y. Y. (2017). Robust object tracking via key patch sparse representation, *IEEE Trans. on Cybernatics* **47**(2), 354–364.

Hinami, R., Mei, T. and Satoh, S. (2017). Joint detection and recounting of abnormal events by learning deep generic knowledge, *The IEEE International Conference on Computer Vision (ICCV)*.

Hsueh, Y.-L., Lie, W.-N. and Guo, G.-Y. (2020). Human behavior recognition from multiview videos, *Information Sciences* **517**, 275–296.

Ke, Y., Sukthankar, R. and Hebert, M. (2007). Event detection in crowded videos, *2007 IEEE 11th International Conference on Computer Vision*.

Kwon, J. and Lee, K. m. (2014). Tracking by sampling and integrating multiple trackers, *IEEE Tran. on PAMI* **36**(7), 1428–1441.

Li, X., Shen, C., Dick, A. R., Zhang, Z. M. and Zhuang, Y. (2016). Online metric-weighted linear representations for robust visual tracking, *IEEE Trans. Pattern Anal. Mach. Intell.* **38**(5), 931–950.

Liang, J., Jiang, L., Niebles, J. C., Hauptmann, A. G. and Fei-Fei, L. (2019). Peeking into the future: Predicting future person activities and locations in videos, *IEEE Conference on Computer Vision and Pattern Recognition (CVPR)*.

Mademlis, I., Tefas, A. and Pitas, I. (2018). A salient dictionary learning framework for activity video summarization via key-frame extraction, *Information Sciences, Elsevier* **432**, 319–331.

Milbich, T., Bautista, M., Sutter, E. and Ommer, B. (2017). Understanding videos, constructing plots learning a visually grounded storyline model from annotated videos, *IEEE ICCV*, Venice, Italy, pp. 4404–4414.

Pal, S. K. and Chakraborty, D. B. (2017). Granular flow graph, adaptive rough rule generation and tracking, *IEEE Trans. Cybernatics* **47**(12), 4096–4107.

Pawlak, Z. (1992). *Rough Sets: Theoretical Aspects of Reasoning about Data*, Kluwer Academic Publishers, Norwell, MA.

PETS-2001 (2001). *IEEE Int. WS Perfor. Evaluation of Tracking and Surveillance.*

Possegger, H., Sternig, S., Mauthner, T., Roth, P. M. and Bischof, H. (2013). Robust real-time tracking of multiple objects by volumetric mass densities, *IEEE Proc. on CVPR*.

Song, S. and Xiao, J. (2013). Tracking revisited using rgbd camera: Unified benchmark and baselines, *Proceedings of IEEE ICCV*, IEEE, Washington, DC, USA, pp. 233–240.

Stauffer, C. and Grimson, W. E. L. (2000). Learning patterns of activity using real-time tracking, *IEEE Trans. on PAMI* **22**, 747–757.

Wang, L., Lu, H. and Yang, M.-H. (2018). Constrained superpixel tracking, *IEEE Trans. on Cybernatics* **48**(3), 1030–1041.

Yang, Y., Liu, J. and Shah, M. (2009). Video scene understanding using multi-scale analysis, *IEEE Intl. Conf. on Comp. Vision*, Kyoto, pp. 1669–1676.

Zaidenberg, S., Boulay, B. and Bremond, F. (2013). A generic framework for video understanding applied to group behavior recognition, *IEEE AVSS*, Beijing, pp. 136–142.

Zhang, K., Liu, Q., Wu, Y. and Yang, M.-H. (2016). Robust visual tracking via convolutional networks without training, *IEEE Trans. Image Proc.* **25**(4), 1779–1792.

Index